Canned Heat

Ethics, Human Rights and Global Political Thought

Series Editors: Sebastiano Maffettone and Aakash Singh Rathore
Center for Ethics & Global Politics, Luiss University, Rome

Whereas the interrelation of ethics and political thought has been recognized since the dawn of political reflection, over the last sixty years — roughly since the United Nation's Universal Declaration of Human Rights — we have witnessed a particularly turbulent process of globalizing the coverage and application of that interrelation. At the very instant the decolonized globe consolidated the universality of the sovereign nation-state, that sovereignty — and the political thought that grounded it — was eroded and outstripped, not as in eras past, by imperial conquest and instruments of war, but rather by instruments of peace (charters, declarations, treaties, conventions), and instruments of commerce and communication (multinational enterprises, international media, global aviation and transport, internet technologies).

Has political theory kept apace with global political realities? Can ethical reflection illuminate the murky challenges of real global politics?

This Routledge Book series *Ethics, Human Rights and Global Political Thought* addresses these crucial questions by bringing together outstanding monographs and anthologies that deal with the intersection of normative theorizing and political realities with a global focus. Treating diverse topics by means of interdisciplinary techniques — including ethics and applied philosophy, political theory, international relations and human rights theories, international political economy, and theories of globalization, including postcolonial studies — the books in the Series presents up-to-date research that is accessible, practical, yet scholarly. These volumes will prove of great relevance to researchers, educators and students, as well as politicians, policy-makers and government officials.

Also in the Series

Wronging Rights? Philosophical Challenges for Human Rights
Editors: Aakash Singh Rathore and Alex Cistelecan
ISBN 978-0-415-61529-7

Conflict Society and Peacebuilding: Comparative Perspectives
Editors: Raffaele Marchetti and Nathalie Tocci
ISBN 978-0-415-68563-4

Global Justice: Critical Perspectives
Editors: Sebastiano Maffettone and Aakash Singh Rathore
ISBN 978-0-415-53505-2

Deprovincializing Habermas: Global Perspectives
Editor: Tom Bailey
ISBN 978-0-415-85933-2

Between Ethics and Politics: Gandhi Today
Editor: Eva Pföstl
ISBN 978-0-415-71064-0

Canned Heat

Ethics and Politics of Global Climate Change

Editors

Marcello Di Paola
Gianfranco Pellegrino

Routledge
Taylor & Francis Group
LONDON NEW YORK NEW DELHI

First published 2014 in India
by Routledge
912 Tolstoy House, 15–17 Tolstoy Marg, Connaught Place,
New Delhi 110 001

Simultaneously published in the UK
by Routledge
2 Park Square, Milton Park, Abingdon, Oxon OX14 4RN

Routledge is an imprint of the Taylor & Francis Group, an informa business

Typeset by
Solution Graphics
A–14, Indira Puri, Loni Road
Ghaziabad, Uttar Pradesh 201 102

Printed and bound in India by
Avantika Printers Private Limited
194/2, Ramesh Market, Garhi, East of Kailash
New Delhi 110 065

British Library Cataloguing-in-Publication Data
A catalogue record of this book is available from the British Library

ISBN 978-1-138-02027-6

Contents

Acknowledgements

The editors wish to thank Aakash Singh Rathore and Sebastiano Maffettone, who included this volume in their series *Ethics, Human Rights, and Global Political Thought*. We also thank David Held and Dale Jamieson, not just for contributing to, but also for supporting this project from the start. We are grateful to all contributors who made our job easy and a privileged occasion for learning. We also wish to thank the Center for Ethics and Global Politics at LUISS, which provided consistent support throughout. We are grateful to Springer for their kind permission to reprint parts of Dale Jamieson's 'Climate Change, Responsibility, and Justice', included in *Science and Engineering Ethics* 16(3): 431–45.

Introduction
The Ethics and Politics of Climate Change: Many Themes, a Common Global Challenge

Marcello Di Paola and *Gianfranco Pellegrino*

According to recent data, the ice-sheet that covers Greenland is getting thinner.[1] This is one outcome of the so-called greenhouse effect: increased concentrations of certain gases in the atmosphere — mainly carbon dioxide, nitrous oxide and methane — cause increased ratios of solar radiation to remain entrapped within the atmosphere itself. Air warmed by the greenhouse effect melts the top surface of the ice-sheet; the resulting water leaks through fissures, lubricating the slip and speeding the seaward flow of glaciers. The warmer water of the sea breaks up the ice, further accelerating the melting process.

Since the 1970s, the average area of sea that remains frozen during summers has decreased by about one-third. The volume of sea ice during the summer months has also decreased, and even faster — because the Arctic ice cap on the whole is thinning. Soon there may be no more ice left at the North Pole. As John Broome recently intimated,

> a few decades from now, our rich, colourful planet will display to a traveller in space only one white polar ice cap instead of two. There could hardly be a more potent symbol of what human beings are doing to our climate than the destruction of one of the Earth's ice caps. It should teach us how large the unforeseen consequences of our acts can be. Since its cause is principally our burning of fossil fuel, it should make us fear what might be the next result of continuing the same way. But avarice has overcome fear. The surrounding nations see the retreat of ice as an opportunity to extract from beneath the Arctic Ocean yet more supplies of oil and gas to burn. They are competing with each

[1] See data from the Polar Science Center, University of Washington, http://psc.apl.washington.edu/wordpress/research/projects/arctic-sea-ice-volume-anomaly/ (accessed 20 April 2013).

other for territorial rights, and already sending out rigs to drill in those dangerous waters (Broome 2012: 2).

Global climate change is the main challenge humanity shall face in the 21st century and beyond. An excessive rise in average global temperature will have significant and long-lasting repercussions not just on North Pole glaciers, but on all those eco-systems that have been making human life and its flourishing possible for thousands of years. Such challenge is self-inflicted: the current rise of average global temperatures is anthropogenic, that is, brought about by us — incarnated by an unusually fast-paced loss of species, more erratic weather patterns, the already mentioned polar ice meltdowns, consequent sea levels rise, and more. In the past 200 years, things have precipitated: the two industrial revolutions have paved the way for the development of patterns of production and consumption that have since then required the releasing of billions of tons of greenhouse gases into the atmosphere. We also know that further effects of climate change are already stored-up for the future, given the atmospheric longevity of those greenhouse gases molecules we have already emitted. It is as if we canned heat.

Today, the question that worries many policy-makers, academics, environmentalists, and lay observers is not *whether*, but *how much* will the earth's climate be changing in the future. The magnitude and rate of climate change is a function of the amount of greenhouse gases that we pump into the atmosphere. The larger the amount of present and future emissions produced by humans, the more significant the climatic and ecological changes that will ensue. Air temperatures, ocean acidity, soil composition, precipitation patterns, and habitat distributions will all be heavily affected by a changing global climate. Extreme weather events such as typhoons, floods and droughts will become more frequent. There is also the possibility of climatic and ecological feedbacks being set off, and tipping points being surpassed, which would cause natural systems to be altered in essential, unpredictable and irreversible ways.[2]

There is a good chance that climate change will impose great burdens and suffering on most, if not all humans (See IPCC 2007: 8–12 and 48–52). This conjures up a scenario of great urgency.

---

[2] On the science of climate change, see Broome (2012: ch. 2) and IPCC (2007).

Unfortunately, this is not reflected in most recent exercises of global governance on the topic. States have thus far failed to come up with agreements and institutions capable of providing a global orientation to climate policy, often for reasons of sheer national interest. As Dale Jamieson once put it, climate change has been a most formidable test for the appropriateness of realist theories of international relations — a test which these theories have brilliantly passed.[3]

However, it remains unclear whether national interests alone will be able to pull the cart. One reason is that the interests in question touch not just on the amount of emissions that countries should be allowed to produce, but on a wide variety of other major economical, political and social dimensions of contemporary life. How many of us there are, what we eat, the way we dwell, the way we move around and transport things, the ways in which we produce goods, and the sort of goods we consume, want and like — these are all basic dimensions of human existence that every state, and hopefully one day all states in concert, must come to confront if they are to take efficacious action against global climate change. Clearly, then, the issue goes way beyond agreeing on some limits to emissions: at stake is the overall feasibility of our current political, economic and cultural infrastructures. In fact, the current rise in average global temperatures could be looked at as a proxy, an indicator of the unsustainability of many of the ways in which we have chosen to live.

In discussing climate change we are confronted with a whole range of issues of great theoretical and practical urgency: this makes it a focal topic for scholarly reflection. Another reason for dedicated scholarly attention is that the problem is a monumental epitome of a whole class of global quandaries. According to Held and McGrew, globalization amounts to the intensification of worldwide social relations and interactions, such that distant events acquire very localized impacts and vice-versa (2003: 3). This definition is both very general and very precise: it can describe and explain phenomena as diverse as nuclear proliferation, financial insecurity, labour market fluctuations and, of course, ecological degradation and climate change in particular. What makes all these phenomena

---

[3] See http://www.philostv.com/dale-jamieson-and-jay-odenbaugh/ (accessed 20 April 2013).

structurally similar is the fact that their sources and impacts are fragmented in space and time and, as a consequence, responsibility for them is a planetary issue: the human race as a whole is responsible for the globalization phenomena. These factors have in most cases led to governance failures — conflicts, financial insecurities, the search for cheaper labour, and climate change have not been dealt with effectively by our current global institutions, if they have been dealt with at all.

Such phenomena also incarnate various asymmetries of power: the status quo that engenders them is one in which vulnerability is unjustly distributed. The global poor as well as future generations — having done nothing to cause climate change, financial insecurity and nuclear proliferation — will nonetheless be most strongly affected by their pernicious reverberations. A relative (if not total) lack of concern for their fate colours current governance inefficacy with sinister shades of malice and calculated unfairness. Discussing climate change, therefore, will also lead us to discuss the structure of many pressing global issues, not just the contingent features of a specific one.

What sets climate change apart, however, is the fact that we lack not just robust institutions to confront it but also robust theories to understand it. In fact, we do not even know exactly how to conceptualize the issue; and that reverberates on how we go about it when making policy.

The Manifold Dimensions of the Climate Change Debate

If an appropriate management of the climate change phenomenon is premised on the clarification, if not the configuration, of the very way in which we should think about it, then a demand arises and a space opens for theoretical investigation. All sciences, natural and social, will have to engage climate change from a multiplicity of angles. And because the phenomenon has also to do with the conceptions of the good life we entertain, with the dignity and well-being of billions of present and future people, the integrity of non-human species and eco-systems, and with the ways in which we organize our ways of living together — then a demand arises and a space opens for a specifically ethical, moral and political theory of climate change. The task is impervious, as we have little by way of fundamentals. As S. Gardiner notes, existing theories are largely underdeveloped in many relevant areas, from intergenerational obligations, and what is good for populations rather than individuals, to international and intra-species justice; from scientific uncertainty and probabilistic thinking, to individual ethics in a global, systemically interconnected world; from what actually constitutes human well-being, to what political participation may mean and achieve when it comes to global matters (2011: 7).

The reason for these theoretical shortcomings is not just our complacency with established modes of thought. It is our actual difficulty in reforming or revising them in the face of drastically changed circumstances. Our motivational sets, moral systems and political institutions have not evolved to engage with global conditions whose sources and impacts are dispersed across wide spatiotemporal stretches. They have rather evolved in relatively small, localized communities, and in response to problems that did not involve future people, or at least not those in the further future. So we lack not just theory, but also reliable intuitions to guide the first steps to its configuration (see Jamieson 2007; see also Mulgan, Ch. 3, in this volume).

Once philosophical reflection sets in, the most remote theoretical complexities of climate change emerge. A list of the many philosophical questions involved would include at least the following: can we apply existing moral theories to such an unprecedented phenomenon, or do we need whole new theories? If it is the latter, to what extent can we rely on the moral intuitions that we already have? If not much, what will be the reference points for configuring these new theories? A parallel worry arises in the political realm: if nation-states and current international institutions have failed to address the problem so far, according to what principles should the structure of the world order be revised? On a more operational plane, what would be the best avenues for the new world order to pursue in fighting climate change: command and control, a cap-and-trade system, de-growth, population management, and/or geo-engineering? These are only few of the many possible trajectories, each with their own political agendas and each commanding that funds be channelled one way rather than another. Which of these, or what mix of which should we choose? And also, for the sake of whom, or what? What, if any, are in fact our obligations towards spatiotemporally distant humans and non-human species or eco-systems? In an inter-temporal framework, are these obligations affected by the fact that by picking one policy over another we may determine not just the levels of well-being of those who shall come into existence, but also who these people will be? The same question arises also with regard to non-human species and eco-systems. Are policies that tend to preserve the integrity of eco-systems to be generally preferred to those that imply significant manipulation and modification, and why? If humans are a species among species, what counts as 'natural' and what doesn't? Does that even matter in the face of possible catastrophes, including the extinction of the human species itself? Is human extinction an especially bad thing? Does the possibility of catastrophic outcomes pose yet further challenges to our moral and political theories? While talking possibilities, what are we to think of in the face of radical uncertainty? And what would, in the case of climate change, be an efficacious interface between science, politics and markets? There are also metaphysical questions regarding the level at which causation is to be attributed — are our individual actions also causes of this planetary phenomenon, or only those of governments and businesses? Who is responsible for climate change? Can we draw clear divides between individual and institutional obligations? To what extent can character-building and education serve us well against climate change in the long term? What is to be considered a virtue in a globally interconnected and hotter world? And how is it to be developed and exercised? Can education and advocacy remain distinct when it comes to such threatening phenomena?

Overview and Aims of the Present Volume

The queries discussed earlier are only some among the conceptual attractors generated by climate change. Quite evidently, they span over a multiplicity of philosophical disciplines and are bound to engender very different theoretical perspectives. In order to protect and further promote such variety, the present volume brings together a range of views from eminent scholars currently engaged in various aspects of the debate. The overall aim is of elucidating climate change as a general philosophical problem, at the same time representing the wide variety of sub-themes it engenders, and the many different theoretical perspectives from which these may be approached.

The volume opens with pieces by David Held, Dale Jamieson and Tim Mulgan. All three provide penetrating diagnosis of the political and moral intricacies of climate change. Held presents the phenomenon as a most acute symptom of a general failure of global governance. Among the structural sources of that failure, he individuates the overall inadequacy of an outdated multilateral order, designed — in the aftermath of the Second World War — as a set of institutions primarily geared to averting political violence among states; as well as the fact that the extension and intensity of the climate change problem sharply increases the governance capacity required, while a strong tension still exists between global interests and state sovereignty. Another set of problems relates to a few structural features of democratic rule (short-termism and interest-group concentration, among others). However, Held argues that no bases exist for abandoning democracy in favour of some form of eco-authoritarianism, rather, a publicly justifiable process of suitable 'will formation' is in order, and the most promising way to implement it is to enhance the deliberative character of democratic decision-making.

Jamieson deepens our understanding of climate change as a radically novel problem, explaining how its very nature and strategic structure challenge the basic tenets of our moral and political systems. Inadequate views of what constitutes a moral wrong and of what political systems ought to take care of stand in the way of our proper comprehension of the problem. In order to conceptualize climate change as clearly involving moral wrongs and global injustices, we will first need to reconsider and revise some central concepts in moral and political theory. Jamieson goes on to argue that climate change threatens another basic environmental value — respect for nature — which cannot easily be taken up by concerns for global justice or moral responsibility. Therefore, the very task of alleviating climate change must awake us to a different way of looking at nature and our place in it, thus leading to a new vision of the role of mankind in a much larger picture.

Mulgan pushes this suggestion, and the related issues, a step further. An ethic for climate change requires us to give up a historical methodological assumption in moral and political philosophy, that is, reliance on pre-theoretical intuitions.[4] What seems firm in standard cases begins to vacillate when familiar scenarios (such as

[4] Cappelen (2012) has recently challenged the idea that any philosopher genuinely made use of this method ever. He claims that many philosophers voice an endorsement of this method, but do not really employ it.

the trolley problem or the drowning child case) are complicated by placing among their background conditions and features magnified spatiotemporal distances, identity-determining choices, and the like. Accordingly, Mulgan suggests, the role for moral theory, and particularly for those moral theories most able to give practical guidance even in the absence of firmly held intuitions, widens. Climate change requires more theorizing as well as new theoretical tools. The world after climate change is a broken world — in which some of the assumptions we held in the last centuries are doomed to vanish. Established ideas such as equality and protection of human rights may reveal themselves as inapplicable in a world made strikingly different by the worst effects of climate change. Perhaps, we should begin to consider abandoning them. In fact, they may even turn out to be morally illegitimate, as they may condemn future people to a life worse than ours.

The second cluster of essays deals with topics in the political theory of climate change governance. Sandler devotes his chapter to a comparative assessment of adaptation and mitigation strategies against climate change. While recognizing the need for and the urgency of adaptation, he reminds us of the importance of not losing sight of the ethics of climate change mitigation. He then presents a case for why climate change mitigation is ethically preferable to adaptation — on grounds of human rights, global justice, environmental value, and welfare. Finally, he proposes several criteria for ethical evaluation of climate change mitigation strategies and policies.

Moellendorf discusses the issue of how to assign responsibility for mitigating and financing adaptation to climate change. He argues that accounts of responsibility, which are tied to the historical responsibility of states, are morally inadequate and defends the application of the ability-to-pay principle. Among the strengths of this conception of responsibility, he individuates a nice fit with the right to sustainable development that poorer countries should enjoy, a fact that climate diplomacy cannot ignore.

Gupta addresses two questions of pressing importance: how legitimate is it to continue to expect and demand leadership from developed countries in a changing global context, characterized by the increasingly strong influence of emerging countries such as India and China? And how can litigation and/or dispute resolution on climate change assist in creating globally relevant legal precedents, which may help bring the climate change negotiation process back on track and stronger than before?

The third cluster of essays is devoted to topics in the moral theory of climate change. Suikkanen and Sahni apply contractualist and virtue theory, respectively, to the phenomenon. Suikkanen sketches an imaginary climate change scenario and suggests moral obligations that decision-makers would have in a contractualist framework. He explains that contractualism would recommend policies that no party involved could reasonably reject, by comparing pair-wise the burdensome consequences of various alternatives. These policies would then set the moral standards by which relevantly similar cases are to be decided. In particular, Suikkanen argues that we will act wrongly unless we take drastic mitigation measures against climate change, because we could not justify our behaviour to all future generations on grounds which they could not reasonably reject.

Sahni argues that early Buddhism contains some responses to two questions of great importance, both relating to the role of the individual in the face of climate change — 'am I to blame?' and 'what ought I to do?' She addresses the first question through the theory of *kamma* (*karma*), and the second by referring to Buddhist virtue theory, indicating that the practice of specific virtues may be conducive to a better understanding and a better fulfilment of the role of the individual in the face of this phenomenon. To an extent, her inspection of Buddhist kamma seems to provide some articulation of that 'respect for nature' that Jamieson had previously introduced in his piece.

Di Paola addresses a similar question, but from a completely different perspective. He rehearses arguments in favour of the idea that individuals are subject to an obligation to engage in self-starting anti-climate change practices, and indicates what sort of practices individuals would have to engage in for that obligation to be appropriately discharged. He argues that such practices cannot be purely private, but must rather be capable of being interpersonally co-ordinated and, when so, of effectively stimulating systemic reform. They must allow for effective signalling and learning; be compatible with individual moral psychology (that is, they must generate individual benefits, not just burdens); and demand of individuals no more than what their measure of personal responsibility justifies. To maximize efficacy (while further diminishing the need for and costs of governmental coercion), they should also be capable of moulding our habits of mind and action in environmentally congenial ways. Di Paola proposes food-producing urban and peri-urban gardening as an example of an appropriate practice.

Orsi follows a different line of research, linking climate with population ethics. He examines the so-called intuition of neutrality, as discussed and refuted by Broome (2004). The intuition tells us that the addition of people does not, by itself, produce or subtract value from the world. Such intuition allows us to disregard the effects — assumed to be ethically neutral — of climate change policy onto the size of populations, effectively allowing us to make policy recommendations. Broome famously argued that the intuition has to go, which immensely complicates the ethics of climate change. Orsi responds to Broome's objections to the intuition by urging a normative (rather than Broome's axiological) interpretation of neutrality, and arguing in favour of an exclusionary permission to disregard the value of adding lives. He explores justifications and limits of such permission by referring specifically to the prospect of human extinction.

The final cluster of essays is devoted to specific ramifications of the climate change problem. Gruen and Loo examine the impact greenhouse gas emissions (arising from industrial agricultural practices) will have on food security, if climate change predictions are right. They argue that a just solution to the potential problem requires a re-conceptualization of responsibility, and urge for a re-evaluation of the strategies to mitigate food insecurity that rely on further agricultural industrialization as, in all likelihood, they will do more harm than good. According to them, individual consumers as well as corporations and governments who have profited from increased food availability — resulting from greenhouse gas intensive farming practices, beneficial changes in climate, or both — have a moral responsibility to aid those who will suffer when food becomes too expensive or unavailable, even if they aren't, in a narrow sense, causally responsible for climate change.

Pellegrino focuses on the predicament of environmental migrants displaced by climate-induced adverse events. He puts forward two claims. First, those people should be given the status of climate, or environmental, refugees, and this is no strain to the regulations and the spirit of the Geneva Convention on the Status of Refugees

(1951). Their claim to the status of refugees arises from their being endowed with a specific right that their states failed to protect — that is, the right to a territory. Second, this individual right to a territory puts a limit to states' right to territory. Accordingly, states have the duty to admit environmental refugees and to give them a settled territory to live in.

Schlottmann discusses climate change education. He presents it as education about and for responding to climate change. He focuses particularly on the ethical purpose of education for responding to climate change. He maintains that such education can include both general (first-order) and specific (second-order) aims, each with different ethical qualities. He notes that education for responding to climate change, especially if it aspires to second-order aims, raises ethical concerns about advocacy and selective information. He discusses the permissibility of advocacy and the role of both science and urgency in climate change education.

The volume closes with a venture in environmental aesthetics. Ciccarelli argues that the occurrence of climate change undermines all attempts of moving from the acknowledgement of nature's aesthetic value to the configuration of moral duties towards it. She notes that some likely outcomes of climate change can, in fact, be aesthetically acceptable, even 'beautiful', by current standards of natural aesthetics (see Ch. 14). The conclusion is that when it comes to climate change, the aesthetic foundation of ecological preservationism is fatally compromised. Ciccarelli further maintains that the aesthetic appreciation of climate change requires the adoption of a 'cosmic' perspective, that is, an evaluative standpoint able to negotiate geological time.

2014 Update: Nothing New — Data from the Fifth IPCC Assessment Report

Readers of this book will soon grasp that climate change is an 'impure' subject-matter: it mixes politics, science, economics, morality, and epistemology, and often the blend amounts to an entangled heap of difficult puzzles, tentative hypotheses and contested assumptions. Climate change is the only topic in contemporary politics about which people quarrel not only on substantive political solutions, but even on the very fact that the problem at stake is really occurring. As many readers might know, sceptics deny the very fact that dangerous anthropogenic climate change is obtaining and will continue to obtain in the next centuries.[5] Climate change is periodically subjected to a scientific as well as political assessment in specialized reports and boards, the most famous being the one provided by the Intergovernmental Panel on Climate Change (IPCC) — whose fifth version was published on 25 September 2013, after the draft of this book had been completed. Accordingly, this book relies on data from the penultimate report, issued in 2007. Surprisingly, though, this implies no flaw in the argumentative line and the factual bases of this book, for the new report does not alleviate our worries. Rather, it confirms the danger that a business-as-usual strategy engenders for humankind and future generations. And in this case, no news is not good news.

The fifth report provides confirmation of the most negative predictions on the following specific issues:

(*a*) Degrees of certainty: according to the authors of the report, 'warming of the climate system is unequivocal'. In particular, 'each of the last three decades has been successively warmer at the Earth's surface than any preceding decade since 1850' and '1983–2012 was likely the warmest 30-year period of the last 1400 years'. Over the period 1880–2012, the combined temperature of land and temperature warmed of 0.85C°. Finally, 'it is virtually certain that the upper ocean (0–700 m) warmed from 1971 to 2010' (IPCC 2013, Sec. B, B.1. and B.2).

(*b*) Increase of extreme weather and climate events: this has been observed since the 1950s — 'the number of cold days and nights has decreased and the number of warm days and nights has increased', as well as 'the frequency of heat waves in large parts of Europe, Asia and Australia ... The frequency or intensity of heavy precipitation events is likely to be increased in North America and Europe' (ibid., Sec. B.1).

(*c*) Increased rate of glacier loss: 'the average rate of ice loss from the Greenland ice sheet has very likely substantially increased' of a fourth more over the period 2002–11. The Antarctic ice sheet has been subjected to an increase of one-third in the same period (ibid., Sec. B.3).

[5] On climate change scepticism, see Oreskes and Conway (2010: chs 6 and 7).

(*d*) Increased rate of sea level rise: 'it is very likely that the mean rate of global averaged sea level rise was 1.7 mm/yr between 1901 and 2010, 2.0 mm/yr between 1971 and 2010, and 3.2. mm/yr between 1993 and 2010' (ibid., Sec. B.4).

(*e*) Climate change is a human-induced phenomenon: 'human influence on the climate system is clear.' It is evident 'from the increasing greenhouse gas concentrations in the atmosphere, positive radiative forcing, observed warming, and understanding of the climate system'. Indeed, 'it is extremely likely that human influence has been the dominant cause of the observed warming since the mid-20th century' (ibid., Sec. D and D.3).

(*f*) Projected impacts and predictions: IPCC projected changes into the future by using several climate models (ibid., Sec. D.1). Across these scenarios, global surface temperature change for the end of the 21st century (2081–2100) is likely to range from 2 to 4°C (see ibid., 2013, Sec. E.1). In the same period, the projected sea level rise will range from 50 to 80 cm (ibid., Sec. E.6), global glacier volume will further decrease (ibid., Sec. E.5), and precipitations will increase as well (ibid., Sec. E.2).

Considering these points, we can say that the factual data of the present volume are, in fact, a bit optimistic. Had the most recent data been considered, the factual grounds of the proposals and reflections presented in the following pages would have been even bleaker. In any case, the normative substance of the ideas given here is unchanged. But the urgency of the invocations and the relevance of the ethical and political remarks voiced are even more pressing today, than when this manuscript was presented to the publisher.

References

Broome, John. 2012. *Climate Matters: Ethics in a Warming World*, W.W. Norton & Company, New York.

———. 2004. *Weighing Lives*, Oxford University Press, Oxford.

Cappelen, Herman. 2012. *Philosophy without Intuitions*, Oxford University Press, Oxford.

Gardiner, Stephen. 2011. *A Perfect Moral Storm: The Ethical Tragedy of Climate Change*, Oxford University Press, Oxford.

Held, David and Anthony McGrew (eds). 2003. *The Global Transformation Reader: An Introduction to the Globalization Debate*, Polity Press, Cambridge.

Intergovernmental Panel on Climate Change (IPCC). 2013. 'Summary for Policymakers', in Thomas F. Stocker, Dahe Qin, Gian-Kasper Plattner, Melinda M.B. Tignor et al. (eds), *Climate Change 2013: The Physical Science Basis*. Contribution of Working Group I to the Fifth Assessment Report of the Intergovernmental Panel on Climate Change, Cambridge University Press, Cambridge.

———. 2007. *Climate Change 2007: Synthesis Report*, IPCC, Geneva.

Jamieson, Dale. 2007. 'The Moral and Political Challenges of Climate Change', in Susanne C. Moser and Lisa Dilling (eds), *Creating a Climate for Change. Communicating Climate Change and Facilitating Social Change*, Cambridge University Press, Cambridge, 475–82.

Oreskes N. and Erik M. Conway (eds). 2010. *Merchants of Doubt: How a Handful of Scientists Obscured the Truth on Issues from Tobacco Smoke to Global Warming*, Bloomsbury Press, New York.

Part I

Basic Themes: Governance, Morality and the Role of Theory

1

Climate Change, Global Governance and Democracy: Some Questions

*David Held**

The paradox of our times can be stated simply — the collective issues we must grapple with are of growing cross-border extensity and intensity and, yet, the means for addressing these are weak and incomplete. Complex global processes, from the ecological to the financial, connect the fate of communities to each other across the world, yet the problem-solving capacity of the global system is in many areas not effective, accountable, or fast enough to resolve current global challenges (Hale et al. 2013). While there are a variety of reasons for the existence of these problems, at the most basic level the persistence of the paradox remains a problem of governance. In our increasingly interconnected world, no solution to global problems can be achieved by any one nation-state acting alone. Global challenges call for collective and collaborative action — something that the nations of the world have not been good at, and which they need to be much better at if these pressing issues are to be adequately tackled. The abilities of states to address critical issues at the regional and global level are handicapped by a number of structural difficulties, domestic and international, which compound the problems of generating and implementing urgent policies with respect to global goods and bads.

In particular, insufficient progress has been made in creating a sustainable framework for the management of climate change. The failure to generate a sound and effective framework for managing global climate change is one of the most serious indications of the challenges facing the multilateral order. Against the backdrop of 9/11, former British chief scientist David King has warned

* Parts of this article are adapted revisions from David Held and Angus F. Hervey, 'Democracy, Climate Change and Global Governance', 2009, http://www.policy-network.net/publications_detail.aspx?ID-3406 (accessed 22 April 2013).

that the threat posed by climate change is more serious than that of terrorism (2004: 177) and Sir Nicholas Stern has referred to it as 'the greatest market failure the world has ever seen' (2006: xviii). In the broad view of the scientific community, climate change has the capacity to wreak havoc on the world's diverse species, biosystems and socio-economic fabric, and the process has clearly begun.

The Limits of Global Governance

While complex global processes connect the fate of communities to each other across the world, global governance capacity is under pressure, for two reasons. First, the multilateral order, founded after the Second World War, was designed in a different era and, above all, as a set of institutions to help prevent a Third World War and consider when, and under what circumstances, war might be legitimate. Climate change does not fit readily into these priorities. Second, the extensity and intensity of the challenge of globalization sharply increases the nature of the governance capacity required.

The difficulties faced by international agencies and organizations stem from many sources including the tension between universal values and state sovereignty, built into them from their beginning. Many global political and legal developments since 1945 do not just curtail sovereignty, but support it in many ways. From the United Nations (UN) Charter to the 1992 Rio Declaration on the Environment onwards, international agreements often serve to entrench the international power structure. The division of the globe into powerful nation-states, with distinctive sets of geopolitical interests, was embedded in the articles and statutes of leading international governmental organizations (Held 1995: chs 5 and 6). Thus, the sovereign rights and prerogatives of states are frequently affirmed alongside more universal principles.

Further, the reach of contemporary regional and international law rarely comes with a commitment to establish institutions with the resources and authority to make declared universal rules, values and objectives effective. The susceptibility of the UN to the agendas of the most powerful states, the partiality of many of its enforcement operations (or lack of them altogether), the underfunding of its organizations, the continued dependency of its programmes on financial support from a few major states, and the weaknesses of the policing of many environmental regimes (regional and global) are all indicative of the disjuncture between universal principles (and aspirations) and their partial and one-sided application. Three additional deep-rooted problems need highlighting (Held 2004: ch. 6).

First, a set of problems emerges as a result of the development of globalization itself, which generates public policy problems that span the 'domestic' and the 'foreign', and the interstate order with its clear political boundaries and lines of responsibility. There is a fundamental lack of ownership of many problems at the global level. It is far from clear which global public issues are the responsibilities of which international agencies. There is no clear division of labour among the myriad international governmental agencies; functions often overlap, mandates frequently conflict and aims and objectives too get blurred. There are a number of competing and overlapping organizations and institutions, all of which have some stake in shaping different sectors of global public policy. The institutional fragmentation and competition leads not just to the problem of intersecting jurisdictions among agencies, but also to the problem of issues falling between agencies. This latter problem is especially manifest between the global level and national governments.

A second set of difficulties relates to the inertia found in the system of international agencies — the inability of these agencies to mount collective problem-solving solutions, faced with uncertainty about lines of responsibility and frequent disagreements over objectives, means and costs. This often leads to a situation where the cost of inaction is greater than the cost of taking action. The failure to act decisively in the face of urgent global problems not only compounds the costs of dealing with these problems in the long-run, but it can also reinforce a widespread perception that these agencies are not only ineffective but unaccountable and unjust as well.

A third set relates to a democratic deficit, itself linked to two interconnected problems — the power imbalances among states as well as those between state and non-state actors in the shaping and making of global public policy (Held 2004). Multilateral bodies need to be fully representative of the states involved in them but they rarely are — having a seat at the negotiating table at a major international governmental organization or a conference does not ensure effective representation. For, even if there is parity of formal representation (a condition often lacking), it is generally the case that developed countries have large delegations equipped with extensive negotiating and technical expertise, while poorer developing countries frequently depend on one-person delegations; they even have to rely on sharing a delegate.

Underlying these institutional difficulties is the breakdown of symmetry and congruence between decision-makers and decision-takers (Held 1995: part I), which would mean that those who are significantly affected by a global good or bad should have a say in its provision or regulation — that is, the span of a good's benefits and costs should be matched by the span of the jurisdiction in which decisions are taken about that good (Held 2004: 97–101). Yet, all too often, there is a collapse — between decision-makers and decision-takers, between decision-makers and stakeholders, and between the inputs and outputs of the decision-making process.

The Limits of Democratic Politics

The urgent challenge of climate change also poses a critical test for modern democracy. While at the level of global governance there has been a failure to generate an effective international framework for managing climate change, state level solutions are usually weak and struggle to transcend the normal push and pull of partisan politics. Modern liberal democracies suffer from a number of structural characteristics that weaken their capacity to tackle global collective action problems in general and climate change in particular. These are:

Short-termism

The electoral cycle tends to focus policy debate on short-term political gains and satisfying the median voter. The short duration of electoral cycles ensures that politicians are concerned with their own re-election, which may compromise hard policy decisions that require a great deal of political capital. It is extremely difficult for governments to impose large-scale changes on an electorate, whose votes they depend on, in order to tackle a problem whose impact will largely be felt by future generations.

Self-referring Decision-making

Democratic theory and politics builds on a notion of accountability linked to domestic constituencies. It assumes a symmetry and congruence between decision-makers and decision-takers within the boundaries of the nation-state. Any breakdown of equivalence between these parties tends not to be heavily weighed. Democratic 'princes' and 'princesses' owe their support to that most virtuous source of power — their people. The externalities or border spillover effects of the decisions they take are not their primary concern.

Interest Group Concentration

In democracies, greater interest group pluralism reduces the provision of public goods because politicians are forced to adopt policies that cater to the narrow interests of small, well-organized groups (Olson 1982). The democratic process rewards these groups and, consequently, it leads to their proliferation. In addition, strong competition among such groups leads to gridlock in public decision-making, delaying both the implementation and effectiveness of public goods provision (Midlarsky 1998).

Weak Multilateralism

Governments accountable to democratic publics often seek to avoid compliance with binding multilateral decisions if this weakens their relationship to their electorate (although there is a notable exception that occurs when strong democratic governments are able to control the multilateral game).

Concerns such as these have generated scepticism about the compatibility of democratic forms of governance with the need for drastic and urgent changes in policy required to combat climate change. The implication is that democracies are unable to meet the scale of the challenge posed by the phenomenon and that more coercive forms of government may be necessary. Such thinking finds its historical precedent in the work of the 'eco-authoritarians' of the 1970s, who argued that it might be difficult for democracies to constrain economic activity and population growth that results in pressures on the environment. They suggested that some aspects of democratic rule would have to be sacrificed to achieve sustainable future outcomes, since authoritarian regimes are not required to pay as much attention to citizens' rights in order to establish effective policy in key areas (Hardin 1968; Heilbroner 1974; Ophuls 1977).

Democracy versus Autocracy

This type of argument has, however, been challenged by a body of theory arguing that there are a number of reasons why democracies, despite structural limitations, are more likely than authoritarian regimes to protect environmental quality (Holden 2002). Democracies have better access to information, with fewer restrictions on media, and greater transparency in decision-making procedures. They encourage the advance of science, which is responsible for our awareness about climate change and other forms of environmental threat in the first place (Giddens 2008: 74). Scientists and other experts are free to engage in research, exchange new evidence and travel to and obtain information from other countries. These factors make it more likely that environmental issues will be identified and placed on the political agenda as well as tackled according to appropriate measures of risk.

Moreover, concerned citizens can influence political outcomes not only through the ballot box, but through pressure groups, social movements and the free media — channels that are closed in autocracies. The presence of civil society also serves to inform the public, act as a watchdog on public agencies, and directly lobby government (Payne 1995). There are many examples of cases where environmental interest groups have been able to overwhelm business, pursuing environmentally damaging practices, and where they have changed the public agenda (Bernauer and Caduff 2004; Falkner 2007).

At the same time, authoritarian regimes have fewer incentives to adopt or stick to sustainable policies. Environmental concerns are often trumped by economic development plans and external security, as was the case with the Soviet regime (Porritt 1984). Leaders are unaccountable to the public and have less reason to enact long-term policy (Congleton 1992). Furthermore, those in power control a substantial fraction of society's resources, encouraging payoffs to the relatively small elite, resulting in less public goods provision (Bueno de Mesquita et al. 2003).

It does not seem unreasonable, then, to expect a strong correlation between democracy and environmental quality. Indeed, among the 40 highest carbon emitters internationally (cumulatively responsible for 91 per cent of total world emissions), countries that have the best records are all democracies. However, upon closer examination, the record is less compelling, and detailed empirical evidence is inconclusive. Environmental quality is not just measured by a broad-based commitment to addressing emissions of carbon and other greenhouse gases. While some studies have shown that authoritarian regimes have worse records than democracies on environmental protection (See Jancar-Webster 1993 and Desai 1998), others find no evidence to suggest that this is the case (cf. Grafton and Knowles 2004). Indeed, across a range of measures and geographical areas, numerous studies prove that outcomes are varied.[1]

On balance, while evidence on the link between political institutions and environmental sustainability does seem to suggest that democracies are preferable to authoritarian regimes, we might expect the effect to be far greater than it actually is. Why is this the case? Part of the reason might be attributed to the different types of transmission mechanisms that translate policy commitment into policy outcomes. Baettig and Bernauer (2009), for example, find

[1] Midlarsky (1998) finds that democracies have a good record on land area protection, but not on deforestation, carbon dioxide emissions and soil erosion, while Didia (1997) holds that democratic countries in the tropics have lower deforestation rates. Bhattarai and Hammig (2001) claim a similar result in Latin America and Africa. Li and Reuveny (2006) show a positive effect for democracy on emissions, deforestation, land degradation, and water pollution but Barrett and Grady (2000) find that while political and civil freedoms mostly impact positively on air pollution, results for water pollution are mixed. Torras and Boyce (1998) maintain that democracy is statistically insignificant for dissolved oxygen, fecal coliform and particulates emissions. Neumayer (2002) demonstrates that democracies sign more multilateral environmental treaties and comply more fully with international obligations while Ward (2008) claims that liberal democracies generally promote sustainability in fossil fuel emissions, but only very weakly.

that while the effect of democracy on political commitment to climate change is positive, the effect on policy outcomes, measured in terms of emissions and trends, is ambiguous. They observe that the causal chain from environmental risks to public perceptions of such risks, to public demand for risk mitigation, and to policy output is shorter than the one leading from risk via policy output to policy outcome. As a result, outcomes are influenced by a range of other factors, such as the properties of the resource in question, mitigation costs and the efficiency of implementing agencies. Politicians might easily declare a set of public policy commitments to climate change mitigation but the outcome of such efforts is affected by factors that are often outside of their control. The result is that policy-makers respond quite well to public demands for more environmental protection but tend to discount implementation problems, hoping that voters will not be able to identify these within a short enough time period to use their votes as a punishment for any failure to deliver.

Political Commitment and the Deliberative Democracy Approach

Political commitment to tackling climate change is critical, yet may require political leaders to adhere to a particular course of action that is potentially unpopular and, hence, contrary to structural democratic pressures. The actual implementation of policies that reduce global warming may infringe on the democratic preferences of citizens. In such a context, political leaders can be caught between a desire for recognition and esteem in the international community — recognition that comes from peer admiration for leadership — and the need to ensure accountability to domestic electorates (Keohane and Raustiala 2008). However, good democratic leadership is not confined to policy-making alone — it also involves educating constituents about pressing issues that may not always be obvious. In this sense, the fact that democratic publics do not always have fully-formed preferences is an advantage as well as a risk. Citizens can significantly shift their preferences, faced with new information and evidence about pressing issues. The democratic citizen that is capable of being 'fact-regarding, future-regarding and other-regarding' is not simply a myth (Offe and Preuss 1991: 156–57).

Such an approach to democratic 'will formation' can be found within the tradition of what is known as deliberative democracy, broadly defined as 'any one of a family of views according to which the public deliberation of free and equal citizens is the core of legitimate political decision-making and self-governance' (Bohman 1998: 401). Deliberative democrats advocate that democracy moves away from any notion of fixed and given preferences — to be replaced with a view that democracy should become a learning process in and through which people come to terms with the range of issues they need to understand in order to hold defensible positions. They argue that no set of values or particular perspectives can lay claim to being correct and valid by themselves but rather are valid only in so far as they are capable of public justification (Offe and Preuss 1991: 168). Individual points of view need to be tested in and through social encounters that take into account the point of view of others. Ultimately, the key objective is the transformation of private preferences through a process of deliberation into positions that can withstand public scrutiny and test. Empirical findings show that citizens can and do alter their preferences when they engage with new information, fresh evidence and debate (Held 2006: 247–55). These can lead to new and innovative ideas about public policy and about how democracy might function and work.

Deliberative democracy can, in principle, increase the quality, legitimacy and, therefore, the sustainability of environmental policy decisions. This is partly due to the uncertainty associated with environmental issues, which demands a wide range of experience, expertise and consultation. The complexity of climate change problems also requires integrated solutions that have been vetted by multiple actors, which cut across the narrow confines of expert knowledge and the responsibilities of established institutions and organizations. And, the concerns of environmental justice require political process to be as inclusive as possible, giving voice to the under-represented including future generations. Effective and just action on climate change depends upon the continuing involvement of citizens in the making and delivery of policy; conventional representative democracy is a poor way to achieve this alone. To remodel environmental politics around deliberative democracy is thus to create an opening for a change in the way democracies address environmental management, especially climate change.

It will not be easy to achieve this deliberative reframing of liberal democracy. The question now is whether democratic systems can be evolved to handle environmental issues better, and how this may be achieved. But certainly the alternative — in the form of authoritarian regimes — is a much worse prospect. The way ahead seems clear even if the journey will be fraught with difficulties.

References

Baettig, Michèle B. and Thomas Bernauer. 2009. 'National Institutions and Global Public Goods: Are Democracies More Cooperative in Climate Change Policy?' *International Organization* 63(2): 281–308.

Barrett, Scott and Kathryn Grady. 2000. 'Freedom, Growth, and the Environment', *Environment and Development Economics* 5(4): 433–56.

Bernauer, Thomas and Ladina Caduff. 2004. 'In Whose Interest? — Pressure Group Politics, Economic Competition and Environmental Regulation', *Journal of Public Policy* 24(1): 99–125.

Bhattarai, Madhusudan and Michael Hammig. 2001. 'Institutions and the Environmental Kuznets Curve for Deforestation: A Cross-country Analysis for Latin America, Africa and Asia', *World Development* 29(6): 995–1010.

Bohman, James. 1998. 'The Coming of Age of Deliberative Democracy', *The Journal of Political Philosophy* 6(4): 400–25.

Bueno de Mesquita, Bruce, Alastair Smith, Randolph Siverson, and James Morrow. 2003. *The Logic of Political Survival*, MIT Press, Cambridge.

Congleton, Roger. 1992. 'Political Institutions and Pollution Control', *Review of Economics and Statistics* 74(3): 412–21.

Desai, Uday. 1998. 'Environment, Economic Growth, and Government', in Uday Desai (ed.), *Ecological Policy and Politics in Developing Countries*, State University of New York Press, Albany, NY, 1–46.

Didia, Dal O. 1997. 'Democracies, Political Instability and Tropical Deforestation', *Global Environmental Change* 7(1): 63–76.

Falkner, Robert. 2007. *Business Power and Conflict in International Environmental Politics*, Palgrave Macmillan, Basingstoke.

Giddens, Anthony. 2008. *The Politics of Climate Change*, Polity Press, Cambridge.

Grafton, Quentin and Stephen Knowles. 2004. 'Social Capital and National Environmental Performance: A Cross-sectional Analysis', *Journal of Environment and Development* 13(4): 336–70.

Hale, Thomas, David Held and Kevin Young. 2013. *Gridlock: Why Global Cooperation is Failing*, Polity Press, Cambridge.

Hardin, Garrett. 1968. 'The Tragedy of the Commons', *Science* 162(3859): 1243–48.

Heilbroner, Robert L. 1974. *Inquiry into the Human Prospect*, Norton, New York.

Held, David. 2006. *Models of Democracy*, Polity Press, Cambridge.

———. 2004. *Global Covenant*, Polity Press, Cambridge.

———. 1995. *Democracy and the Global Order: From the Modern State to Cosmopolitan Governance*, Polity Press, Cambridge.

Holden, Barry. 2002. *Democracy and Global Warming*, Continuum, London.

Jancar-Webster, Barbara. 1993. 'Eastern Europe and the Former Soviet Union', in Sheldon Kamieniecki (ed.), *Environmental Politics in the International Arena: Movements, Parties, Organisations and Policy*, State University of New York Press, Albany, 199–222.

Keohane, Robert, and Kal Raustiala. 2008. 'Toward a Post-Kyoto Climate Change Architecture: A Political Analysis', *UCLA School of Law, Law and Economics Research Paper Series*, Research Paper no. 08–14.

King, David. 2004. 'Climate Change Science: Adapt, Mitigate, or Ignore?' *Science* 303(5655): 176–77.

Li, Quan and Rafael Reuveny. 2006. 'Democracy and Environmental Degradation', *International Studies Quarterly* 50(4): 935–56.

Midlarsky, Manus. 1998. 'Democracy and the Environment: An Empirical Assessment', *Journal of Peace Research* 35(3): 341–61.

Neumayer, Eric. 2002. 'Do Democracies Exhibit Stronger International Environmental Commitment? A Cross-country Analysis', *Journal of Peace Research* 39(2): 139–64.

Offe, Claus and Ulrich Preuss. 1991. 'Democratic Institutions and Moral Resources', in David Held (ed.), *Political Theory Today*, Polity Press, Cambridge, 143–71.

Olson, Mancur. 1982. *The Rise and Decline of Nations*, Yale University Press, New Haven.

Ophuls, William. 1977. *Ecology and the Politics of Scarcity*, Freeman, San Francisco.

Payne, Rodger. 1995. 'Freedom and the Environment', *Journal of Democracy*, 6(3): 41–55.

Porritt, Jonathan. 1984. *Seeing Green: The Politics of Ecology Explained*, Basil Blackwell, Oxford.

Stern, Nicholas. 2006. *The Economics of Climate Change: The Stern Review*, Cambridge University Press, Cambridge.

Torras, Mariano and James Boyce. 1998. 'Income, Inequality, and Pollution: An Assessment of the Environmental Kuznets Curve', *Ecological Economics* 25(2): 147–60.
Ward, Hugh. 2008. 'Liberal Democracy and Sustainability', *Environmental Politics* 17(3): 386–409.

2

Climate Change, Responsibility and Justice

Dale Jamieson

I begin by characterizing the kinds of risk that climate change poses, and go on to describe several distinct kinds of practical responsibilities which it is widely thought to engage. After discussing these kinds of practical responsibilities in some detail, I briefly discuss the value of respect for nature. Finally, I draw some conclusions.

Risk and Responsibility

Climate change poses two different kinds of risk that can roughly be characterized in two ways. First, it poses the risk of a large, rapid, relatively linear change; and second, climate change poses the risk of an even larger, more rapid, non-linear change. The first sort of risk is typically discussed in the climate impacts literature and generally presupposed by economic models. While it is sometimes noted that climate change poses other sorts of risk, most models and assessments — especially those carried out by economists, lawyers and policy professionals — focus on risks of this first kind.[1] The second kind is more frequently highlighted by scientists, environmental advocacy groups and (alas) Hollywood filmmakers.[2]

These different kinds of risk pose many of the same questions because they mostly have the same causes. However, they also pose some different questions because the first kind of risk is more likely to produce some relatively predictable winners, at least to some limit (say 2°C), than the second kind of risk, which is more likely to produce only losers and be less predictable, both in its occurrence and in its effects.

---

[1] For the claim that much work in climate change economics fails to address the second sort of risk adequately, see Weitzman (2007).

[2] For a study that focuses on this second sort of risk, see the National Academy of Science Report (2002).

Since these risks are being imposed by human action, at least to a great extent, they raise questions of what I call 'practical responsibility' that can be distinguished from both theoretical and causal responsibility, though practical responsibility may be closely associated with both. Practical responsibility concerns what we are responsible for doing while theoretical responsibility concerns what we are responsible for thinking. Exactly what relationships obtain between practical and theoretical responsibilities is dependent on the association between doing and thinking and between practical and theoretical reason generally. Practical and theoretical responsibility can both be distinguished from causal responsibility in that the former notions are normative, while it is widely believed that the latter concept is fundamentally descriptive.[3]

I take practical responsibility to include varieties of both prudential and ethical responsibility. Prudential responsibility centres on responsibilities that one has to oneself while ethical responsibility centres on responsibilities that one has to others. This otherwise (fairly) clear distinction can be muddled by shifting the identity of the agent who bears the responsibility. If I have some specific responsibility to my family (in the sense in which I am using the terms) it is an ethical responsibility; however, if the family (of which I am a part) has the same responsibility to itself, then the family has a prudential responsibility and, perhaps, so do I in virtue of being a member of the family. I leave it as an open question as to whether either prudential or ethical responsibility should be seen as a variety of the other or as falling under some more general category.

It seems reasonably clear that being causally efficacious is a necessary condition for being practically responsible — for imposing, reducing or eliminating a risk.[4] However, I will leave it as an open question whether counterfactual causal efficacy can, in some cases,

---

[3] However, recent work by Joshua Knobe (2006) suggests that attributions of causal responsibility often follow attributions of practical responsibility rather than the other way around.

[4] Jules Coleman may be denying this when he claims that a case in which 'you've done nothing about shovelling the snow from your walkways' and someone 'coming to visit you slips and breaks her leg . . . is a case of responsibility without causation' (1992: 274). For present purposes, I will leave aside the question of whether he is denying my claim, and if so, whether his claim would be plausible.

satisfy this condition. I will also leave it open as to what may be the other conditions for being practically responsible.

People are causally efficacious in different ways and respects. For example, I am causally efficacious when I choose the vegetarian option and when I organize an animal rights group. I am also causally efficacious in my role as a consumer and as a professor. Since agents are causally efficacious, through individual and collective action as well as through institutional roles, it is plausible to suppose that practical responsibility is plural and layered.

Practical responsibility is plural in that the same act can discharge multiple practical responsibilities — for example, when I keep my promise to another person by participating in a political demonstration. Practical responsibility is layered in that agents can have practical responsibilities at different levels of social organization where they are casually efficacious — for example, as individual voters and as members of a political party (for more information, see Jamieson 2005).

My claim is that climate change engages several distinct kinds of practical responsibility: prudential and ethical, with the ethical including the moral and political. However, each of these responsibilities, while figuring in how we ought to respond to climate change, deviates from standard cases in which those kinds of responsibility obtain.

Prudential Responsibility

Prudential reasons for responding to climate change may seem quite strong, especially when one reflects on the fact that conservative politicians, such as Margaret Thatcher, Nicolas Sarkozy and Angela Merkel, have forcefully advocated action on this issue. The prudential case for acting on climate change is often made by analogy to arguments for buying insurance. For example, Stephen Schneider has said:

> A continuation of 'business as usual' raises a serious concern from the risk-management point of view, given that the likelihood of warming beyond a few degrees before the end of this century (and its associated impacts) is a better than even bet. Few security agencies, businesses or health establishments would accept such high odds of potentially dangerous outcomes without implementing hedging strategies to protect themselves, societies and Nature from the risks of climate change in our case. This is just a planetary scale extension of the risk-averse principles

> that lead to investments in insurance, deterrence, precautionary health services and business strategies to minimise downside risks of uncertainty (Schneider 2005: 15726).

Schneider has summarized this point in the following way: 'we buy fire insurance for our house and health insurance for our bodies. We need planetary sustainability insurance' (ibid.).[5]

I agree that our present policy is contrary to most reasonable notions of enlightened self-interest and that there is some force to the insurance analogy. However, there are also some important disanalogies between investing in climate protection and purchasing insurance.

First, we have no actuarial tables for the climate protection market in the way that we do for accidents and fires. We have very little idea (much less any notion of statistical reliability) about the specific impacts of climate change on societies like ours, living on terrestrial planets; much less do we have any data about how specific changes in the composition of the atmosphere are likely to have these impacts. The disanalogies run even deeper. Insurance is typically purchased by an agent to benefit her/himself or, in some cases, those whom s/he loves or to whom s/he feels responsible. But in this case, we would be asking people who are now living very well, who (under many scenarios) have adequate resources for adaptation, to buy insurance that will mainly benefit poor people, who will live in the future in some other country; and to do this, primarily, on the basis of predictions about the future based on climate models, expert reports and so on. Moreover, rich people do not for the most part love or feel responsible for their poor contemporaries, especially those who live across national boundaries, much less for those who will live in the future.

Another way of making the case for collective prudential responsibility is through an economic assessment of the aggregate expected damages of climate change and the costs of avoiding them. This approach views the human community as a single agent and compares the aggregate costs and benefits of various policies.

---

[5] Weitzman (2007) also employs the insurance analogy. While Jared Diamond (2005) does not employ this analogy, his work has become a popular *locus classicus* for the view that we have prudential reasons to be concerned about environmental degradation.

There are reasons for doubting that the case for responding aggressively to climate change can be made simply in terms of economics. First, the human community is not a single agent acting on the basis of rational self-interest. Human communities are diverse, involving individuals with different interests, and are not (in the economists' sense) perfectly rational or even, in many cases, aspiring to be so. Any climate change will have distributional effects and the model of humanity as a single agent cannot adequately reflect such distributional conflicts. Second, viewing climate change as a problem of prudential responsibility typically involves treating all preferences as commensurable, usually by monetarizing them. But diverse values are at stake, such as biodiversity protection and social solidarity, in addition to economic values, such as income and assets. It is not clear that all such values can be meaningfully placed on the same scale, much less monetarized. Finally, the idea that one can know enough to perform reliable damage assessments of climate change, up to the end of the century and beyond, is patently absurd. One does not have to be a genius to recognize that we can barely predict the state of the economy from one quarter to another.

Moral Responsibility

It is often said that climate change is a matter of individual moral responsibility. The climate change issue can be seen at its core as centring on rich people appropriating more than their share of a global public good and, in addition, harming poor people by causally contributing to extreme climatic events, such as droughts, hurricanes and heat waves. Moreover, much of this behaviour is unnecessary, even for maintaining the profligate lifestyles of the global rich (see Shue 1993). As plausible as this is, once we begin to model the problem of climate change on more familiar cases of individual moral responsibility, important differences begin to emerge.

There are various paradigms of what constitutes a moral problem but the following sort of case is surely at the centre: an individual acting intentionally harms another individual; both the individuals and the harm are identifiable; and the individuals and the harm are closely related in time and space. Consider Example 1, the case of Jack intentionally stealing Jill's bicycle.[6] The individual acting

[6] I first introduced this series of examples in Jamieson (2007b).

intentionally has harmed another individual, the individuals and the harm are clearly identifiable, and they are closely related in time and space.

If we vary the case on any of these dimensions, we may still see the case as posing a moral problem, but its claim to be a paradigm moral problem weakens. Consider some further examples. In Example 2, Jack is part of an unacquainted group of strangers, each of which, acting independently, takes one part of Jill's bicycle, resulting in the bicycle's disappearance. In Example 3, Jack takes one part from each of a large number of bicycles, one of which belongs to Jill. In Example 4, Jack and Jill live on different continents and the loss of Jill's bicycle is the consequence of a causal chain that begins with Jack ordering a used bicycle at a shop. In Example 5, Jack lives many centuries before Jill and consumes materials that are essential to bicycle manufacturing; as a result, it will not be possible for Jill to have a bicycle. While it may still seem that moral considerations are at stake in each of these cases, this is less clear than in Example 1, the paradigm case with which we began.

The view that morality is involved is weaker still, perhaps disappearing altogether for some people, if we vary the case on all these dimensions at once. Consider Example 6: acting independently, Jack and a large number of unacquainted people set in motion a chain of events that causes a large number of future people, who will live in another part of the world, from ever having bicycles. For some people, the perception persists that this case poses a moral problem. This is because it may be thought that the core of what constitutes a moral problem remains. Some people have acted in a way that harms other people. However, most of what typically accompanies this core has disappeared. In this case, it is difficult to identify the agents and the victims or the causal nexus that obtains between them; thus, it is difficult for the network of moral concepts (for example, responsibility, blame and so forth) to gain traction.

These 'thought experiments' help to explain why many people do not see climate change as a moral problem, or if they do see it as a moral problem, it fails to have the urgency of a paradigm moral problem. Structurally, climate change is most analogous to Example 6. A diffuse group of people is now setting in motion forces that will harm a diffuse group of future people. Indeed, if anything, the harm caused by climate change will be much greater than the loss of the opportunity to have a bicycle. Still, we tend not to conceptualize this as an urgent moral problem because it is not accompanied by the characteristics of a paradigm moral problem.[7] Climate change is not a matter of a clearly identifiable individual acting intentionally so as to inflict an identifiable harm on another identifiable individual, closely related in time and space.

There are other paradigms of moral responsibility but, as a first approximation, they do not seem to apply neatly to climate change either. For example, in cases of strict liability we hold an agent responsible for the consequences of an action even if there was no malign intention. Much of the argument for strict liability rests on economic and policy considerations but, historically, strict liability was also often applied to acts that were regarded as especially dangerous, such as storing explosives. However dangerous it may be for us all to drive to the store, this rationale does not seem to apply to any one of us driving to the store. In other cases, we hold someone responsible even if they have no malign intent, if they are negligent in failing to act in a way that satisfies a standard of reasonable care. Perhaps, a case could be made that present and future high emitters of greenhouse gases are negligent in this way, but it is not easy to make this case when it is widely believed that human action is not a primary cause of the climate change that is now underway.[8]

There is a deeper problem about whether contributing to climate change is a matter of moral responsibility. The paradigm that I have been discussing views the causation of harm as being at the centre of what makes an act a matter of moral concern. Even if harm causation is neither necessary nor sufficient for an act or omission to be of moral concern, that some such connection exists has been

[7] I am assuming that the perception of urgency flags as a problem drifts further from the paradigm. Whether or not this is true is worthy of further investigation.

[8] According to recent reports, 44 per cent of American voters say that climate change is primarily caused by long-term planetary trends rather than human activity. It could be argued that these Americans are culpable in their ignorance of the relation between human action and climate change, but when prominent public figures are climate change deniers and science education is so obviously inadequate, it is difficult to make this case.

a very influential, if not universally shared, view in modern moral philosophy.[9]

However, recent work in social psychology suggests that when it comes to construing an act or omission as within the domain of morality, other considerations are just as important to people as harm causation. Jonathan Haidt and his colleagues have claimed that considerations involving fairness and reciprocity, in-group and loyalty, authority and respect, and purity and sanctity are (in addition to considerations about the causation of harm) at the foundation of morality as conceived by most people.[10] Since these considerations can come apart, often people will deny that harm-causing activity is within the moral domain, while at the same time considering behaviour that does not cause harm to be of moral import. Daniel Gilbert (2006) brings these considerations to bear on the question of climate change when he writes that

> global warming doesn't . . . violate our moral sensibilities. It doesn't cause our blood to boil (at least not figuratively) because it doesn't force us to entertain thoughts that we find indecent, impious or repulsive. When people feel insulted or disgusted, they generally do something about it, such as whacking each other over the head, or voting. Moral emotions are the brain's call to action. Although all human societies have moral rules about food and sex, none has a moral rule about atmospheric chemistry. And so we are outraged about every breach of protocol except Kyoto. Yes, global warming is bad, but it doesn't make us feel nauseated or angry or disgraced, and thus we don't feel compelled to rail against it as we do against other momentous threats to our species, such as flag burning. The fact is that if climate change were caused by gay sex, or by the practice of eating kittens, millions of protesters would be massing in the streets.

Climate change, therefore, presents us with an issue that displays some of the marks of a paradigm moral problem but fails to exhibit others. Viewing climate change as a problem of individual

---

[9] The most thorough treatment of the normative significance of harm causation is Joel Feinberg's magisterial four-volume work (Feinberg 1984–88). Though criminal law is Feinberg's main concern, much of what he says applies to morality as well.

[10] For an introduction to this work visit http://people.stern.nyu.edu/jhaidt/home.html (accessed 23 April 2013).

moral responsibility to some extent requires a revision of everyday understandings of moral responsibility.

Political Responsibility

Climate change also seems to present us with another challenge to our notion of ethical responsibility. In addition to being a rather deviant case of individual moral responsibility, it provides us with the political challenge of securing global justice. Most of the emitting is done by the rich countries of the North, but most of the climate-change related dying is done in the poor countries of the South (Patz et al. 2005).

Such facts seem to lead to the conclusion that climate change poses questions of global justice. While this may be true, there are complications. Since the atmosphere does not attend to national boundaries and a molecule of carbon has the same effect on climate wherever it is emitted, climate change is largely caused by rich people and is suffered by poor people, wherever they live. We can attribute primary responsibility for climate change to the 500 million people who emit half of the world's carbon, but not all of them live in the rich countries of the North.[11] While it is difficult to get accurate statistics that would precisely locate these 500 million people, we can use automobile ownership as a rough proxy. As of 2002, there were about 800 million cars in the world with more than 230 million in the United States (US), 76 million in Japan and nearly 50 million in Germany. Further, there were more than 20 million cars in China and more than 17 million in India, while there were only about 18 million in Canada, 12.5 million in Australia, and a little over 2 million in New Zealand (Dargay et al. 2007). It, thus, seems plausible to suppose that more of those people who are the principal causes of climate change live in China than in Canada, Australia and New Zealand (or in many European countries, such as Austria, Belgium, Switzerland, and the Netherlands, for that matter). Moreover, since poor people suffer the most from climate change, wherever they live, it is plausible to suppose that like those who contribute to causing climate change, those who will suffer the most are also distributed around the globe.

I do not want to deny that climate change poses questions of global justice. Rather, my point is that, like questions of individual

---

[11] Here I rely on data from Steve Pacala (personal communication).

moral responsibility, the problems that climate change presents us with stray from the paradigm of global justice. In several important respects, causing climate change is not like one country unjustly invading another country. The nation-state is one level of social organization that is relevant to addressing climate change because it is causally efficacious, but it is not the primary bearer or beneficiary of ethical responsibilities in this regard.

What I have claimed in this as well as the previous section is that we cannot simply say that climate change confronts us with a clear case of ethical responsibility. We can argue (to my mind, plausibly) that climate change does pose questions of ethical responsibility, but this argument would have to be revisionary. We would have to show that there are good reasons for extending or revising our concepts of ethical responsibility in such a way that problems posed by climate change would fall under them.

Respect for Nature

Thus far I have claimed that both prudence and ethics can be seen as providing reasons to respond to climate change, but in both cases they stray from the norm. In my view there is another value that climate change puts at risk, which is often not noticed, and recognizing this helps to explain why some people are so passionate about this issue. I call this value 'respect for nature' and I claim that embracing it should motivate people to acknowledge a responsibility to respond to climate change.[12] While I think that such a duty is recognized by many people, it is difficult to make it comprehensible and to defend. Like many duties, it is easier to say when it is violated than when it is respected.

In 1997, a distinguished group of scientists published an influential article in which they assessed the human impact on nature (Vitousek et al. 1997). They calculated that (*a*) between one-third and one-half of the Earth's land surface had been transformed by human action; (*b*) carbon dioxide in the atmosphere had increased by more than 30 per cent since the beginning of the industrial revolution; (*c*) more nitrogen had been fixed by humanity than all other terrestrial organisms combined; (*d*) more than half of all accessible

[12] I am greatly indebted to Paul Taylor's early and important work on this topic; however, it will become clear that my conception of respect for nature is significantly different from his. See Taylor (1989).

surface fresh water was being appropriated by humanity; and (*e*) about one-quarter of the Earth's bird species had been driven to extinction. From these facts, they inferred that 'it is clear that we live on a human-dominated planet' (Vitousek et al. 1997: 494). It is, of course, apparent that over the last decade these measures of human domination have only increased. While it may be difficult to say what exactly the duty of respect for nature consists in, it seems clear that where there is such a duty, human domination violates it (this much Kant would have agreed with).

Why should we suppose that humans dominate nature? Vitousek and his colleagues provide evidence for the human domination of nature, but the question of what the domination consists in remains open. There is a tradition in environmental ethics that thinks of nature as autonomous and (as is the case for humans) domination is thought to (roughly) consist in undermining autonomy through arbitrary interference.[13] Some will cavil at the idea that nature can be thought of as autonomous, but consider the following. It is not entirely clear what autonomy means, even in the case of humans, but to some extent it seems to relate to being self-caused.[14] If we think of nature as that which is distinct from humanity, then it is clear why Vitousek and his colleagues think that the facts that they report show that humans dominate nature. Rather than being governed by its own laws and internal relations, nature is increasingly affected by human action. Humans, like other forms of life, influence their environments and affect the nature that gave rise to them, but what makes the present human relationship with nature one of domination is the degree and extremity of human influence on nature. At some point, the causal influence is so thorough-going that it can be said to constitute domination.[15]

I will assume in what follows that no conclusive argument has been given that prevents us from saying, along with Vitousek and

---

[13] See, for example, Katz (1997), the essays collected in Heyd (2005) and Turner (1996). What Turner means by 'wildness' is related to what I mean by 'autonomy'. For reservations, see O'Neill et al. (2008: 134–37).

[14] See Jerome Schneewind (1997) who argues that autonomy is a relatively recent conceptual construction with its origins in Kant.

[15] For more on these themes, see Jamieson (2002:190–96, 2008: 166–68).

his associates, that we live on a human-dominated planet and that if there is a duty of respect for nature, then human domination violates that duty. Anthropogenic climate change violates the duty of respect for nature because it is a central expression of the human domination of nature.

This domination is expressed both substantively and attitudinally. The numbers cited earlier show the substantive nature of human domination. It is also expressed attitudinally in the ways that we think about nature and feel about our relations with it. It is not an exaggeration to say that as a civilization we treat the Earth and its fundamental systems as carelessly as if they were toys, as if their functions could easily be replaced by a minor exercise of human ingenuity. It is as if we have scaled up slash-and-burn agriculture to a planetary level.[16] Seen in this way, our collective behaviour towards nature seems to be a paradigm of disrespect.

Thus far I have discussed why we might think that we are violating a duty of respect for nature on condition that we have such a duty. But what can be said in favour of the view that we have such a duty in the first place?

In the subsequent section, I will tentatively explore three possible grounds for supposing that we have such a duty. What I say is speculative, not conclusive. Much more work would have to be done to build a convincing case for the existence of such a duty.

One ground for supposing that there is a duty of respect for nature is prudential. We promote our own interests when we respect nature. Versions of this argument are ubiquitous in environmental literature and something like this view is implicit in slogans such as Barry Commoner's 'third law of ecology' (1971), which states that 'nature knows best' and Wallace Broecker's analogy that emitting greenhouse gases is like poking a dragon with a sharp stick. Something like this view can also be seen as providing the foundation for the precautionary principle. In its crudest and most general form, it is in the background of the claim by Costanza and his colleagues that the minimum value of the services that ecosystems provide is between USD 16 and USD 54 trillion, and that their study 'highlights the relative importance of eco-system services and the potential impacts on our welfare of continuing to

[16] I owe this image to Jeremy Waldron (personal communication).

squander them (1997: 259). How plausible one finds this as a foundation of a duty to respect nature depends on one's views about duties to others founded on prudence, as well as on one's view about the substance of these claims. It is worth noting that many people would find little to object to here.

A second reason for respecting nature is that it provides a background condition for our lives having meaning. While it would be implausible to think that it is a necessary or sufficient condition for all lives having meaning, it does seem to be a very important condition in several cultures at many times. It is easy to think of examples of the contribution of nature to life's meaning from history, literature or contemporary culture. Blake's idea of England as a 'green and pleasant land' is important both in literature and in English history. The cherry orchard in Chekhov's play of the same name defines the life of everyone in the community. Think of the role landscape plays in the cultures of indigenous peoples. What I want to suggest is that nature provides the background against which we live our lives, thus providing us with an important source of meaning. This, it might be claimed, is sufficient for supposing that we have a duty to respect nature. For when we fail to do so, we lose an important source of meaning in our lives.

A third reason for respecting nature is from a concern for psychological integrity and wholeness. As Kant (and later Freud) had observed, respecting the other is central to knowing who we are and to respecting ourselves. Indeed, the failure to respect the other can be seen as a form of narcissism. One can imagine a kind of natural history that views the recognition of nature as the 'other', beyond our control, at the root of self-identity and communal life.[17]

Much more would have to be said to make any of these views plausible or to say what a duty of respect for nature would come to. What I hope to have accomplished in this section is to show that it may be plausible to suppose that there is such a duty, that such a duty need not be based on a morally extravagant view, such as biocentrism or eco-centrism, and that such a duty may be relevant to our climate-destabilizing behaviour.

---

[17] Respecting the otherness of nature can break in at least two different directions: one towards seeing nature as a partner, the other towards aestheticizing nature and seeing it as an object of the experience of the sublime. I say a little more about this in Jamieson (2007a, 2008).

Conclusion

In this article I have discussed the view that the risks of anthropogenic climate change impose practical responsibilities, some of which are prudential and some of which are ethical. I have claimed that while these views are plausible, they would require revising our conceptions of moral and political responsibility. I have also suggested that in addition to the duties generated by these responsibilities, another duty — respect for nature — also seems engaged by the risk of climate change. This duty is not widely discussed and is under-theorised and defended. However, I suspect that unless a duty of respect for nature is widely recognised and acknowledged, there will be little hope of successfully addressing the problem of climate change.

References

Coleman, Jules. 1992. *Risks and Wrongs*, Cambridge University Press, New York.

Commoner, Barry. 1971. *The Closing Circle*, Knopf, New York.

Costanza, Robert, Ralph d'Arge, Rudolf de Groot, Stephen Farber et al. 1997. 'The Value of the World's Ecosystem Services and Natural Capital', *Nature* 387(115): 253–60.

Dargay, Joyce, Dermot Gately and Martin Sommer. 2007. 'Vehicle Ownership and Income Growth, Worldwide: 1960–2030', *Energy Journal* 28(4): 163–90.

Diamond, Jared. 2005. *Collapse: How Societies Choose to Fail or Succeed*, Penguin, London.

Feinberg, Joel. 1984–88. *The Moral Limits of the Criminal Law*, Oxford University Press, New York.

Gilbert, Daniel. 2006. 'If Only Gay Sex Caused Global Warming', *Los Angeles Times*, 2 July.

Heyd, Thomas (ed.). 2005. *Recognizing the Autonomy of Nature: Theory and Practice*, Columbia University Press, New York.

Jamieson, Dale. 2008. *Ethics and the Environment: An Introduction*, Cambridge University Press, Cambridge.

———. 2007a. 'Justice: The Heart of Environmentalism', in Ronald Sandler and Phoedra Pezzullo (eds), *Environmental Justice and Environmentalism: The Social Justice Challenge to the Environmental Movement*, MIT Press, Cambridge, 85–101.

Jamieson, Dale. 2007b. 'The Moral and Political Challenges of Climate Change', in Susanne C. Moser and Lisa Dilling (eds), *Creating a Climate for Change: Communicating Climate Change and Facilitating Social Change,* Cambridge University Press, Cambridge, 475–82.

———. 2005. 'Duties to the Distant: Humanitarian Aid, Development Assistance, and Humanitarian Intervention', *Journal of Ethics* 9(1–2): 151–70.

———. 2002. *Morality's Progress,* Oxford University Press, Oxford.

Katz, Eric. 1997. *Nature as Subject,* Rowman and Littlefield, Lanham.

Knobe, Joshua. 2006. 'The Concept of Intentional Action: A Case-study in the Uses of Folk Psychology', *Philosophical Studies* 130(2): 203–31.

National Academy of Science Report. 2002. *Abrupt Climate Change: Inevitable Surprises,* National Research Council: National Academy Press, Washington.

O'Neill, John, Alan Holland and Andrew Light. 2008. *Environmental Values,* Routledge, London.

Patz, Jonathan A., Diarmid Campbell-Lendrum, Tracet Holloway, and Jonathan A. Foley. 2005. 'Impact of Regional Climate Change on Human Health', *Nature* 438(7066): 310–17.

Schneewind, Jerome. 1997. *The Invention of Autonomy,* Cambridge University Press, New York.

Schneider, Stephen H. 2005. 'We Buy Fire Insurance for a House and Health Insurance for Our Bodies. We Need Planetary Sustainability Insurance', *Proceedings of the National Academy of Sciences of the United States of America* 102(44): 15725–27.

Shue, Henry. 1993. 'Subsistence Emissions and Luxury Emissions', *Law and Policy* 15(1): 39–59.

Taylor, Paul. 1989. *Respect for Nature,* Princeton University Press, Princeton.

Turner, Jack. 1996. *The Abstract Wild,* The University of Arizona Press, Tucson.

Vitousek, Peter M., Harold A. Mooney, Jane Lubchenco, and Jerry M. Melillo. 1997. 'Human Domination of Earth's Ecosystems', *Science* 277(5325): 494–99.

Weitzman, Martin L. 2007. 'A Review of the Stern Review on the Economics of Climate Change', *Journal of Economic Literature* 45(3): 703–24.

3

Theory and Intuitions in a Broken World

Tim Mulgan

The practical moral importance of climate change is obvious. In this chapter, I focus instead on what the prospect of dangerous human-induced climate change might contribute to a perennial topic in moral philosophy: the respective roles of theory and intuition.

Two Simple Tales and Some Complications

We begin with two examples that are very prominent in recent philosophical debates.

(*a*) 'Pond': Walking to work, I pass a child drowning in a pond. No one else is around. I can save the child at some small inconvenience to myself. (Perhaps my new suit will need to be dry-cleaned or I will be late for an important meeting.) What should I do?[1]

(*b*) 'Trolley': Walking to work, I pass an out-of-control trolley. It will crash into six innocent people, killing them all, unless I push a large person in front of the trolley. What should I do?[2]

Almost everyone has strong, confident moral intuitions about 'Pond'. I ought to save the child and there is something very morally wrong with me if I do not. Interestingly, reactions to 'Trolley' are less uniform.[3]

---

[1] This example was introduced into philosophical literature by Singer (1972).

[2] This example was introduced into philosophical literature by Foot (1967) and Thomson (1976).

[3] Not everyone agrees that it is wrong to push the person in front of the trolley. But most people have the strong intuition that this is wrong.

There are two simple ways to create scenarios where our intuitions are less confident. The first is to take a simple tale, and make it more complex or unfamiliar. Consider a series of departures from our two simple tales — beginning with 'Pond'. We can complicate it by imagining that the child is drowning, not in front of me, but on the other side of the world; by imagining not a single child, but a large number of children; by imagining not a single bystander, but many, each of whom could save at least one child; by adding the new information that someone threw the child into the pond, or that she jumped, or that her indolent parents or teachers failed to teach her to swim; or by shifting the tale to the distant future, where my present action may somehow cause an as-yet-not-existent child to fall into an as-yet-uncreated pond.

Once we focus on consequences in the distant future, we can also then make my action identity-determining for that future child — so that she would not exist at all unless I perform the present action that also causes her to fall into her pond. As Derek Parfit (1984: part four) has noted, such different people choices — where my actions determine who will exist in the future — are much more common that we might expect. (Parfit calls this the 'non-identity problem', as the people who exist in different possible futures are not numerically identical to one another.) Following Parfit, many philosophers have argued that, in different people choices, common moral principles often break down.

Consider a simple tale introduced by Parfit (ibid.: 371). We must choose between two energy policies. The first is completely safe. The second is cheaper but riskier — burying nuclear waste where there is no earthquake risk for several centuries, but a significant risk in the distant future. Suppose we choose this risky policy. Many centuries later, an earthquake releases radiation, killing thousands of people.

Our choice seems clearly wrong. But why? Intuitively, we do wrong because we harm those who die. But suppose the two energy policies lead to radically different futures — with different patterns of migration and social interaction. Now take any particular individual killed by the catastrophe. Suppose, the precise chain of events leading to her existence would not have occurred if we had chosen differently — her parents would not have met and might not even have existed themselves. But now it appears we have harmed no one. For, how can someone be harmed by an action without which she would not exist? And, if we harm no one, how can our choice be wrong?[4]

We turn now to our second simple tale. Recent moral philosophy has seen a vast literature devoted to more elaborate versions of 'Trolley'.[5] To bring a whole series of variations together, consider Peter Unger's baroque tale 'The Switches and Skate' designed to reduce the whole 'Trolley' enterprise to absurdity:

> By sheer accident, an empty trolley, no one aboard, is starting to roll down a certain track. Now, if you do nothing about the situation, your first option, then, in a couple of minutes, it will run over and kill six innocents who, through no fault of their own, are trapped down the line. (So, on your first option, you'll let the six die.) On your second option, if you push a remote control button, you'll change the position of a switch-track, switch A, and, before it gets to the six, the trolley will go onto another line, on the left-hand side of switch A's fork. On that line, three other innocents are trapped and, if you change switch A, the trolley will roll over them. (So, on your second option, you'll save six lives and you'll take three.) On your third option, you'll flip a remote control toggle and change the position of another switch, switch B. Then, a very light trolley that's rolling along another track, the Feed Track, will shift onto B's lower fork. As two pretty heavy people are trapped in this light trolley, after going down this lower fork the vehicle won't only collide with the onrushing empty trolley, but, owing to the combined weight of its unwilling passengers, the collision will derail the first trolley and both trolleys will go into an uninhabited area. Still, the two trapped passengers will die in the collision. On the other hand, if you don't change switch B, the lightweight trolley will go along B's upper fork and, then, it will bypass the empty trolley, and its two passengers won't die soon. (So, on your third option, you'll save six lives and you'll take two.) Finally, you have a fourth option: further up the track, near where the trolley's starting to move, there's a path crossing the main track and, on it, there's a very heavy man on roller skates. If you turn a remote control dial, you'll start up the skates, you'll send him in front of the trolley, and he'll be a trolley-stopper. But, the man will be crushed to death by the trolley he then stops. (So, on your

[4] For further discussion and references to the vast literature on this topic, see chapter 1 of Mulgan (2006). On the non-identity problem in particular, see Roberts and Wasserman (2009).

[5] For a taste of the literature on elaborated 'Trolley' cases, see Kamm (2007).

> fourth option, you'll save six lives and you'll take one.) On reflection, you choose this fourth option and, in consequence, the six are prevented from dying (1996: 90).[6]

We could also transform 'Trolley' into a different people choice involving future people — perhaps by adding a button that transports the trolley into the distant future, where it may vaporize some as-yet-not-existent large person who would not otherwise ever exist. And, as with 'Pond', we could then transform 'Trolley' into a different people choice, by ensuring that my present choice is identity-determining for those future people who might be adversely affected.

Consider two new cases, bringing together all these various departures from our two simple tales.

(*a*) 'Super Future Pond': I am walking to work. On my way, I could take a small inconvenient detour to donate a small sum to a land reclamation project in a distant and impoverished land. I am one of many millions who must choose whether to make this sacrifice. If we do not engage in this project, then, over millennia, natural processes will lead to floods, leaving many children drowning in ponds. The land reclamation project has such widespread social impact that our present collective decision is identity-determining for those possible future children. If we do embark on the project, they will never exist. What should *we* do? And, given my knowledge of the likely behaviour of others, what should I do?

(*b*) 'Super Future Trolley': It is a new day but I am still walking to work. I hear on the morning news that a meteorite is on a long-term collision course with the Earth. Unless something is done, it will, in several millennia, impact in a very destructive way, killing many millions of people. On my way to work, I could take a small inconvenient detour to donate a small sum to a new project to send Bruce Willis into space to divert the meteorite. If diverted, the meteorite will land in a less populated area, killing fewer future people.

[6] Unger (1996: 89) helpfully presents a diagram to explain this complex example.

> I am one of many millions who must choose whether to make this sacrifice. Like yesterday's land reclamation project, the meteor-diversion project is identity-determining for those possible future people. If we embark on the project, they will never exist. What should we do? And, given my knowledge of the likely behaviour of others, what should I do?

There is obviously much we could say about these two cases. For my purposes in this chapter, I only need the modest claim that our intuitions are less confident in these more baroque examples than in our original simple tales.[7]

A second way to weaken our intuitions is by varying background conditions — abandoning implicit assumptions that lie behind our common-sense morality. Is 'Pond' an isolated event, or a daily (even hourly) occurrence? How do children get into ponds? Who is responsible for the existence of ponds (or the abundance of trolleys) in the first place? Is it relevant to the evaluation of 'Pond' that I am walking to work through an affluent suburb of a wealthy country in a world where many people have little or nothing to eat? Or suppose I pass the pond while driving to work. Is it relevant to the evaluation of this situation that I drive to work in a society where carbon emissions are endangering the lives of future people?

Suppose our intuitions are less confident in unfamiliar cases or when familiar background assumptions fail to hold. What implications does this have for moral theory?

Moral Theory and Unfamiliarity

The answer lies in the fact that one common defence of moral theory appeals to its ability to provide guidance in cases where intuitions are unavailable, uncertain or unreliable. If we all had confident intuitions about all particular cases, and especially if we always agreed in our considered moral judgements, then moral theory would have no practical role. It would be a purely academic

[7] Not everyone agrees that our intuitions are less confident in baroque cases. For instance, Unger's own discussion of 'The Switches and Skates' begins as follows: 'To this moderately complex case, most react that your conduct was good behaviour.' (1996: 90).

exercise. Many moral philosophers — especially those working in the utilitarian tradition — would regard this as insufficient justification for moral philosophy. They see the ability to offer moral guidance as an essential feature of a successful moral theory.

Although this is controversial territory, I will assume in this article that the ability to offer guidance when intuition runs out does count in favour of a moral theory. If, in such a case, theory A offers advice while theory B remains silent, this counts in favour of A over B. Similarly, if A offers better guidance in such cases than B, then this too counts in A's favour.

We are most sure of our moral intuitions in familiar, realistic situations, involving a small constant group of contemporaries. Conversely, we are less confident about moral judgements in unfamiliar, unrealistic situations where different possible futures contain different as well as large numbers of people. If situations of the latter sort turn out to be more common — or more significant — from what we had previously thought, then this raises the significance of the general enterprise of moral theory. It also improves the comparative credentials of those theories that cope best with unfamiliarity.

Changes in background conditions can increase unfamiliarity in three ways. (*a*) They can make cases we already recognize as unfamiliar more common; (*b*) They can make cases we already recognize as unfamiliar more significant; and (*c*) They can make some cases more unfamiliar.

In the rest of this article, I will argue that climate change can similarly disrupt our intuitions. I begin by explaining my broader research project on ethics for a broken world.

Ethics for a Broken World

My general project explores the philosophical implications of the possibility that dangerous human-induced climate change may produce a broken world, where resources are insufficient to meet everyone's basic needs; where a chaotic climate makes life precarious; where each generation is worse-off than the last; and where our affluent way of life is no longer an option.[8]

---

[8] I have presented material relating to this project in two invited plenary lectures — 'Utilitarianism for a Broken World', Plenary lecture at conference on two centuries of utilitarianism, Université de Rennes, June 2009;

Modern western political philosophy assumes a society enjoying what John Rawls (1971) called favourable conditions — where all basic needs can be met without sacrificing basic liberties. We then debate the distribution of rights and resources beyond these minimum requirements. By contrast, a broken world lacks favourable conditions. Natural resources are insufficient to meet the basic needs of the population — perhaps, due to inadequate supplies of food and drinkable water. A broken world cannot feed itself and the resources of the earth cannot support all human beings. The climate is very unpredictable and extreme weather events are common. Within this harsh global context, there are many locally broken areas — societies that cannot meet their own needs. Some are more broken than others; indeed, some are uninhabitable by human beings.

This is not our world. Humanity currently has the resources to meet everyone's needs. And our society is not locally broken — not in my stark sense. (Indeed, Rawls [1999] argued that virtually all modern societies enjoy favourable conditions.) But nor is the broken world a merely imaginary scenario. A broken world is one possible future — and perhaps the most likely one. And, of course, the longer we do less than we should, the more likely that future becomes.

The threat of a broken world comes from human-induced climate change. Everything about climate change is controversial in public debate. I want to stress the modesty of my empirical assumptions. I claim only that past and present human behaviour may produce something like a broken world at some point in the future. This simple claim is sufficient to motivate our discussion and no one can reasonably be confident that it is false.

In this article, I set uncertainty aside and merely assume that climate change will produce a broken world. The rest of this article explores the implications of this assumption for moral theory.

Intuitions in a broken world

In a broken world, unfamiliar cases are more common, morally salient and unfamiliar than they are in a more affluent world (enjoying favourable conditions). This is principally because, in a broken

---

and 'Ethics for a Broken World', Plenary Lecture at the annual conference of the British Society for Ethical Theory, University of Nottingham, July 2010.

world, a number of key background assumptions of recent political philosophy no longer hold. These include the following:

(*a*) 'The assumption of impotence': we cannot affect distant future people.
(*b*) 'The no conflict assumption': there is no conflict between the interests of present people and those of distant future people.
(*c*) 'Rawls's optimistic assumption': future people will be better-off than us.[9]
(*d*) 'Rawls's (present) favourable conditions assumption': all basic needs can be met without compromising basic liberties.
(*e*) 'Rawls's (future) favourable conditions assumption': favourable conditions will persist indefinitely into the future.
(*f*) 'Rawls's (international) favourable conditions assumptions': the vast majority of societies (*i*) currently enjoy favourable conditions; and (*ii*) will continue to enjoy favourable conditions indefinitely into the future.
(*g*) 'The assumption of autarkism': individuals can survive outside of society and only need to co-operate to advance their mutual interests beyond mere survival.[10]
(*h*) 'The laissez-faire assumption': liberal institutions (such as free markets, democratic government or individual rights) promote the common good, without conflicting with the survival of either the individual or her community.

These related differences between the broken world and the idealized world of liberal political philosophy increase the gap between our common-sense moral intuitions — which presuppose these assumptions — and the broken world.

We consider the impact of the broken world under three headings, corresponding to the three ways that unfamiliar background assumptions can render our moral intuitions less confident.

---

[9] Conditions (c), (d), (e), and (f) are all suggested by the discussion in Rawls (1971: 284–93). For further discussion, see Mulgan (2006: ch. 2).

[10] This assumption is especially prominent in the argument of Gauthier (1986). For further discussion, see Mulgan (2006: ch. 2).

The Unfamiliar is More Common

Our intuitions are less reliable in emergency situations; and they are designed for a world where those situations are comparatively rare. In a broken world, analogues of 'Pond' and 'Trolley' are likely to be common occurrences. And these will not be the simple original tales but the more complex variants, involving many potential helpers and many potential beneficiaries — most of whom are distant future people. Those who dwell in the broken world will regularly confront 'Super Future Pond' and 'Super Future Trolley'.

The Unfamiliar is More Significant

The broken world raises the significance of our obligations to distant future people, by removing three key reasons for setting those people aside: (*a*) the belief that we cannot affect them; (*b*) the belief that there is no real conflict between their interests and ours; and (*c*) the optimistic assumption that future people will inevitably be better-off than ourselves.

Of course, no one believes that our actions have no impact on the future. But some people do believe that, because the foreseeable long-term impacts of different actions negate one another, we can safely set distant future people aside. This complacency collapses once we recognize the significance of long-term environmental effects, such as those that might produce a broken world.

Rawlsian optimists argue that future people are easily accommodated (for further discussion, see Mulgan 2006: ch. 2). We will do the best for future people if we leave behind stable democratic institutions and a thriving economy; these goods outweigh any accompanying environmental bads. Our best interests do not conflict with those of distant future people and we can rest assured that they will be better-off.

For the sake of argument, suppose this complacent optimism is misplaced. What if, in other words, climate change produces a broken future where no one shares Rawls's optimism. In the broken world, moral theories that accommodate obligations to distant future people — and offer advice about intergenerational conflict — will have a very significant prima facie advantage. For instance, I argue elsewhere that consequentialist theories accommodate obligations to distant future people more easily than non-consequentialist theories, especially those based on a social contract. Therefore, the shift to a broken world raises the comparative advantage of consequentialism (see Mulgan 2006, especially chs 2, 7, 8, and 9; 2011).

In a broken world, decisions that impact on distant future people — a paradigm unfamiliar case — are much more significant. And those decisions are most likely to be large-scale collective decisions, not individual one-off choices. This increases the unreliability of our everyday intuitions.

Everything is More Unfamiliar

Finally, the fact that background assumptions fail not only makes unfamiliar cases more common and more significant in a broken world, it also makes them more unfamiliar. Our moral intuitions are designed for a world characterized by Rawls' various assumptions of favourable conditions and intergenerational optimism. We apply our intuitions within a world where emergencies occur against a backdrop of affluence, stability and improvement. In a broken world, where emergencies are the norm and Rawls' assumptions no longer hold, we need different moral rules — and we cannot be confident that our intuitions will locate them for us.

Modern liberals take rights seriously. We regard our rights as non-negotiable, inalienable, inviolable — not to be traded-off against the common good or economic productivity. Rights are barriers against the impersonal utilitarian calculus. Perhaps, we today have good utilitarian reasons to regard the violation of human rights as unthinkable. But this may be a luxury that future people in a broken world will be unable to afford. They will have to rethink the relationship between freedom and survival.

In a broken world, where we cannot all survive, the most urgent moral task is to fairly decide who will survive. We might think of rights not as guaranteeing a worthwhile life but the maximum possible equal chance of such a life. When we are engaged in the part of our theory of justice that attempts to describe ideal political institutions, our focus will then shift from securing needs and liberties for all, to managing a fair distribution of chances to secure those needs and liberties.

Consider a simple case: the allocation of water under conditions of scarcity. If an equal share of water is insufficient for survival, it makes no sense to give everyone an equal inadequate share rather than an equal chance of an adequate share. Similarly, within a broken world, we must allocate scarce chances to survive. The fairest way to allocate any scarce necessity is a lottery. Perhaps, we should first conduct a public deliberation to decide how many people can survive, and then hold a lottery to select them. The knowledge that everyone has an equal chance of survival (or otherwise) would be the focus during the earlier deliberation.

We might christen any social decision procedure to determine who lives and who dies in a survival lottery. To implement such a lottery now, in our comparatively affluent world, would be monstrous. But, in the chaotic climate of the broken world, these survival lotteries might become a regular fact of life. Participating in the lottery, and in the preceding deliberation, may be the best way for future people to develop and express their views about the content of rights — and the limits of rights. Perhaps, for them, this is what a right will be — an equal chance to live or die. And, under this broader definition, survival lotteries are not so unfamiliar. We use similar methods — with a sound utilitarian rationale — when allocating scarce medical resources. We balance fairness and efficiency, and leave some people to die. We may not use a lottery in the literal sense, but we do have ways to decide who lives and who dies.

Utilitarianism is often attacked for its willingness to think the unthinkable. The English Roman Catholic philosopher Elizabeth Anscombe went so far as to describe utilitarian thinking as the product of a corrupt mind (1958: 16–17). In a broken world, where the unthinkable must be thought, this willingness becomes, not a vice, but a necessary virtue.

The shift to a broken world thus reveals the historical contingency of our current notion of rights — and of the intuitions that are based on it. Instead of relying on intuition, we must place our moral theory on a more secure, more utilitarian foundation.

Is Our World Broken?

You may think that, precisely because they live in a broken world, distant future people will have their own moral intuitions — ones that are suited to that world. This may be true. Once the broken world gets going, human moral intuitions may eventually evolve to fit it — although we might still doubt that human intuitions will adjust fast enough to keep pace with rapidly changing conditions. But our intuitions are unreliable in relation to such a world.

So we need moral theory to guide our thinking about the broken world.

But this raises an obvious question: why do we need to think about a broken world at all? After all, we are not living in such a world, as we enjoy favourable conditions. This brings us to the second dimension of my project — the impact of a broken future on us. If we consider only present people and their needs, then our world is not broken. But, on any plausible moral theory, the well-being of future people matters as much as our own. So the needs of 'our world' include the needs of future people, and 'our resources' include its future resources. If, on this wider definition, our resources are insufficient to meet all our needs, if we must choose between present and future needs, then our world is already broken.

The same considerations that will force future people in a broken world to place their future-oriented obligations at the top of the ethical agenda should also lead us to give our obligations to them a similar weight. We need the ethics of the broken world now. We cannot separate abstract discussion of life in the broken world from urgent decisions about how we should live now.

Of course, if we can avoid a broken future at little cost to ourselves, then we should do so. But what if we can only avoid that future by sacrificing the basic needs of some present people — perhaps, through drastically reduced energy use, or by diverting research funding from medicine to agriculture; or perhaps, we need a population reduction so rapid that lives must be lost in the transition.

Take a concrete example. Suppose we discover that, if we insist on 70 years of good health for ourselves and insist on the necessary investment in medical technology, then our descendants can only hope for a reasonable chance of 50 years of moderate health. Can we still regard a lifespan of 70 years as a right? If so, why is it a right for us and not for future people? Or suppose we discover that, while we can guarantee our basic needs, our descendants will need to run a survival lottery. Can we insist on guaranteed survival for ourselves or should we move in their direction — operating a survival lottery across the generations? And, if we did institute an intergenerational survival lottery, what might it look like?

More generally, a broken world may require restrictions on personal liberty on a scale that people have only previously accepted in times of war, or other temporary crisis. Private land might be requisitioned to grow food as might individual labour; the use of fossil fuels for private purposes might be severely curtailed; and individual lifestyle choices — especially reproductive decisions — might be much more tightly regulated and constrained.

Some will reply that we should never sacrifice present liberties — never abandon our hard-won rights — until we are certain that we have to. But this is unrealistic, and self-servingly so. Given the scientific and sociological complexities of climate change, certainty is impossible. If we wait until we are certain, we will never sacrifice anything for anyone. Our moral deliberations must deal in probability, not certainty.

Others will argue that, whatever the consequences, some rights are simply inviolable. The problem is that, faced with a broken future, we cannot help violating some rights. If our way of life is leading to a broken world, then we are now harming future people, we are already violating their rights. If our behaviour might lead to a broken world, then we are being reckless with the lives of future people, which is bad enough. If we cannot change our lifestyles without violating some present rights, then we must choose, not whether to violate rights, but whose rights to violate. And there seems no justification — beyond our own selfishness — for privileging present liberties over future needs.

Because our world is not yet broken for us, we can decide whether to acknowledge that it is broken at all. If we are narrowly rational or if we adopt a theory of justice that reinforces our interests, we will insist on our current rights and leave future people to fend for themselves. Nothing outside ethics can force us to take future people into account. We must recognize the utilitarian foundations of our rights and freedoms, acknowledge their fragile contingency, so that we can begin to re-imagine them before it is too late. Perhaps, we can legitimately give some priority to our own values and rights. But to treat them as inviolable is illegitimately to favour present people over future people — to run a lottery where we all have winning tickets, simply due to an accident of birth. This is neither fair nor efficient. However strongly our intuitions point us in this self-serving direction, any acceptable moral theory must condemn it.

References

Anscombe, Gertrude E. M. 1958. 'Modern Moral Philosophy', *Philosophy* 23(124): 1–19.

Foot, Philippa. 1967. 'The Problem of Abortion and the Doctrine of Double Effect', *Oxford Review* 5: 5–15.

Gauthier, David. 1986. *Morals by Agreement,* Oxford University Press, Oxford.

Kamm, Frances M. 2007. *Intricate Ethics,* Oxford University Press, Oxford.

Mulgan, Tim. 2011. *Ethics for a Broken World: Imagining Philosophy After Catastrophe,* Acumen Press, Durham.

———. 2006. *Future People,* Oxford University Press, Oxford.

Parfit, Derek. 1984. *Reasons and Persons,* Oxford University Press, Oxford.

Rawls, John. 1999. *The Law of Peoples,* Harvard University Press, Cambridge.

———. 1971. *A Theory of Justice,* Harvard University Press, Cambridge.

Roberts, Melinda A. and David T. Wasserman (eds). 2009. *Harming Future Persons: Ethics, Genetics and the Non-identity Problem,* Springer, Dordrecht.

Singer, Peter. 1972. 'Famine, Affluence and Morality', *Philosophy and Public Affairs* 1(1): 229–43.

Thomson, Judith J. 1976. 'Killing, Letting Die, and the Trolley Problem', *The Monist* 59(2): 204–17.

Unger, Peter. 1996. *Living High and Letting Die: Our Illusion of Innocence,* Oxford University Press, Oxford.

Part II

Political Theory of Climate Change Governance

4

The Ethics of Climate Change Mitigation

Ronald Sandler

In this chapter, I present the case for why climate change mitigation is ethically preferable to adaptation — on human rights, global justice, environmental value, and welfarist grounds. I then propose several criteria for ethical evaluation of climate change mitigation strategies and policies.

The Challenge of Mitigation

Climatic and ecological effects from anthropogenic increases in atmospheric greenhouse gases (GHG) concentrations are already occurring (Hansen et al. 2010; NOAA 2011). Moreover, even the complete cessation of GHG emissions would not prevent further changes from happening, given the duration that GHG molecules (particularly carbon dioxide [CO_2]) remain in the atmosphere, the possibility of climatic and ecological feedbacks and tipping points, and climatic and ecological momentum generally (Gillett et al. 2011). At issue now is not whether anthropogenic global climate change will occur but what its magnitude and rate will be, which depends upon the amount of future global GHG emissions. The larger the amount of future emissions, the higher concentrations of atmospheric GHGs will be reached, the greater will be the anthropogenic forcing of the climate system, and the more severe will be the consequent climatic and ecological disruptions — with respect to air temperatures, ocean acidification, precipitation patterns, sea levels, and extreme weather events, for example (IPCC 2007).

The difference in projected atmospheric concentrations of CO_2 and carbon dioxide equivalents (CO_2e) in 2100 between a low future emissions scenario and a high business-as-usual (BAU) emissions scenario was 525 parts per million (ppm) and 945 ppm, respectively. Current atmospheric CO_2 concentrations are approximately

395 ppm. The range of projected increases in global mean surface air temperature between these scenarios is 6.7C° (1.2°C compared to 7.9°C) (Climate Interactive 2011a). Therefore, there remains tremendous potential for reducing the social and ecological challenges and losses associated with global climate change through mitigating future emissions. If a low emissions pathway could be accomplished, there would be far fewer climate change-driven population displacements, species extinctions, economic disruptions, and conflicts, for example. A low emissions path — on which atmospheric concentrations of CO_2 would likely not exceed 450 ppm and global mean surface air temperature increases would likely be limited to 2°C over pre-industrial levels — also is the target to which signatories of the United Nations Framework Convention on Climate Change (UNFCCC) are committed to pursuing, in order to 'prevent dangerous anthropogenic interference with the climate system' (UNFCCC 2009).

Accomplishing a low emissions path will be difficult. It requires reducing global CO_2 emissions from BAU by approximately 6 gigatons (or 6 billion tons) per year by 2020, 77 gigatons per year by 2050, and 101 gigatons per year by 2100, with a cumulative reduction in CO_2 emissions of 6,317 gigatons by 2100 (Climate Interactive 2011b).[1] (To provide a sense of the scope of this challenge in practical terms, one gigaton reduction of CO_2 by 2050 could be accomplished by building approximately 275 nuclear power plants at a cost of approximately USD 2 trillion,[2] and four gigatons could be accomplished by increasing the average fuel efficiency of 2 billion vehicles from 30 mpg to 60 mpg, without increasing the number of miles driven by those vehicles [Pacala and Socolow 2004].) Moreover, this would only accomplish a 'path' to 'possible' climate safety.

Given that current energy use (and associated emissions) are disproportionately attributable to industrialized nations (on a per capita basis), but that future increases in energy demand will occur largely in developing nations, it is practically impossible to achieve

---

[1] Alternative calculations generate somewhat different reduction values, due to variations in emissions data, assumptions about future energy demands, and expected energy source distributions (and so GHG emissions intensities).

[2] Calculated on the basis of Romm (2008).

the mitigation goal of 450 ppm (and 2°C over pre-industrial temperatures) without robust and immediate participation by both industrialized and industrializing nations (US EPA 2010). Thus, accomplishing the low (or climate safety) emissions path would require an enormous, longitudinal global effort.

Mitigation vs Adaptation

Meeting the challenge of mitigation described here would not only be politically and technologically challenging, it may have significant economic and social costs. Substantial resources will need to be committed to developing alternative energy sources and infrastructures, as well as to increasing energy efficiency. Some policies to reduce emissions may slow economic growth and development (at least temporarily), prevent exploitation of some natural resources (for example, timber and oil) and reduce overall levels of consumption (at least, in some places or for some people). Thus, a central policy issue regarding global climate change concerns whether, and to what extent, mitigation or adaptation should be prioritized. There will be economic and social costs associated with responding to global climate change, whether it is done now with mitigation or later with adaptation. The policy question is how to distribute efforts and costs. Should we aggressively mitigate now so that adaptation costs are lower later, or should we do less mitigation now and face higher adaptation costs later? In this section, I argue that several ethical considerations strongly favour aggressive mitigation over adaptation.

The Welfare of Future Generations

One influential approach to addressing whether (and to what extent) to prioritize mitigation or adaption is welfare economics, which aims to identify the distribution of mitigation and adaptation that is socially and economically optimal, that is, which would bring about the best balance of social and economic benefits over costs. Nicholas Stern (2007), for example, has argued in favour of aggressive mitigation on the grounds that it will cost much less socially and economically to mitigate now than it will to adapt later. William Nordhaus (2007a, 2007b), however, has argued that the socially and economically optimal approach would be to begin with some mitigation now (but not nearly so much as Stern advocates) and then increase the intensity of mitigation over time.

Bjorn Lomborg (2001), in contrast to both Stern and Nordhaus, has argued that the opportunity costs of addressing global climate change are sufficiently high that the socially and economically optimal approach would be to use those resources to help the world's worst off in other ways — for example, by providing them with medicine, education and resources for economic development.[3]

One of the primary reasons that welfare economic studies of global climate change produce divergent results regarding mitigation and adaptation is that they use different normative assumptions. For example, in order to make all social and economic costs calculable, welfare (such as pleasure, suffering, subjective well-being, and death) needs to be quantified in monetary metrics. There are no obvious or standard rates for converting, for example, malnutrition or being a refugee into a monetary value; so, there is often divergence with respect to value assignments and conversion rates. However, the normative assumption that primarily drives apart welfare economics calculations regarding global climate change concerns the relative weight placed on present versus future generations.

It is common in economics to increasingly 'discount' the future — that is, to not count a dollar today the same as a dollar tomorrow, but rather to count a dollar today as worth more than a dollar in the future, and still more than a dollar in the further future. One reason for this is the expectation that a dollar today will be worth more than a dollar in the future in terms of purchasing power or goods acquisition. Purchasing power is used in economics as a proxy for the capacity to satisfy preferences. This, in turn, is taken as a measure of (or, in some cases, as being constitutive of) welfare or well-being, which is the ultimate concern of welfare economics. Thus, different economists will use a different discount rate if they have different views about future economic growth or contraction. A second reason for divergence in the discount rates used by welfare economists is that they adopt different pure time preference rates, in which the welfare of future generations is discounted simply because they are in the future. The greater the pure time preference, the less the interests or welfare of future generations

[3] Posner and Sunstein (2007) defend a similar view. In his more recent work, Lomborg (2007) has distanced himself from his earlier view.

is considered (or counted) in comparison to the welfare of present generations. Stern employs a lower pure time preference than Nordhaus. Therefore, future costs and harms carry much more weight in Stern's calculus than they do in Nordhaus'. The more that costs and harms to individuals in future generations matter, the more it makes sense to take on costs now to avoid them later. The harms associated with global climate change are backloaded; they will be greater in the future than they are now. Therefore, calculations on whether to prioritize mitigation or adaption to global climate change, and at what time points to do so, are particularly susceptible to variations in the discount rate.

The extent to which the welfare of future generations should be discounted against the welfare of present generations (if at all) is a normative question that cannot be addressed by economics or any descriptive science or social science alone. It has to do with how future generations ought to be valued. This is a vexing question, since future generations do not currently exist, and this existence is contingent on the actions and policies of present generations. Which future people there will be depends on what present people do, including what they do with respect to global climate change. Different climate change policies and technologies will result in different people meeting under different circumstances and, ultimately, having different children. As a result, individuals in future generations cannot be made worse off by us than they would otherwise be — that is, harmed in the standard sense — since, for any future person, the alternative is not that of being better off, but that of not being at all (non-existence). Nevertheless, future people can come into conditions that are more or less conducive to human flourishing. Moreover, it is possible to compare the welfare of possible future generations.

I will not defend a position on the moral standing of future generations here. But the foregoing is sufficient to establish the following conditional conclusion: to the extent that the welfare of future generations should be considered in policies and practices that will impact them, mitigation of global climate change is ethically preferable to adaptation to global climate change.

Distributive Justice

Wealthy people are disproportionately responsible for GHG emissions (per capita), since they consume more goods and energy than those who are poor.[4] They are also better resourced economically, technologically and socially to adapt to the impacts of global climate change. Climate justice proponents believe that such considerations — for example, regarding historical responsibility, present adaptive capacity and distribution of benefits and burdens more generally — need to inform discussions of who should be responsible for mitigation and adaptation (Baer et al. 2008). However, justice considerations are also relevant to whether to prioritize mitigation over adaptation.

Consideration of distributive justice favours mitigation over adaptation because the effects of mitigation are globally shared, whereas adaptation is locally targeted. GHG emissions reductions decrease overall atmospheric CO_2e concentrations, regardless of where they occur. If large emissions reductions are accomplished in already highly industrialized nations, it reduces the magnitude of global climate change overall, and so has widely distributed benefits. In contrast, adaptation is localized and targeted to particular communities or areas. For example, if a lower emissions path is accomplished, such that sea level rises are limited to around 18 cm by 2100, compared to 56 cm on higher emissions scenarios (IPCC 2007) (or several metres if either the west Antarctic ice sheet or the Greenland ice sheet collapse [Pritchard et al. 2009; Joughin and Alley 2011]), all coastal communities are benefited. However, if North American coastal cities build a dyke system similar to Amsterdam's in order to protect themselves from 56 cm sea level rises, only residents of those cities are benefited.

One way in which emissions mitigation can (and, as argued earlier, must) occur is through development by industrializing nations along a lower emissions pathway. This can be fostered through technology and resource assistance on the part of wealthier industrialized nations. The UNFCCC's Green Fund provides a mechanism for this, as does the clean development mechanism (CDM) and an adaptation fund that are part of the Kyoto Protocol, for

[4] In 2005, the wealthiest 20 per cent of the world were responsible for 76.6 per cent of global consumption and the wealthiest 10 per cent were responsible for 59 per cent of global consumption, whereas the poorest 20 per cent were responsible for 1.5 per cent of global consumption and the poorest 10 per cent for 5 per cent of global consumption (World Bank 2008).

example (UNFCCC 2012a, 2012b). It can also be accomplished by requiring lower emissions intensities (that is, fewer GHG emissions per unit of economic activity or per capita), as development thresholds are reached (Costa et al. 2011). Thus, prioritizing mitigation over adaptation does not require foregoing economic development on the part of industrializing nations, and it does not require that industrializing nations shoulder a larger social burden of mitigation than already highly industrialized nations. Instead, it requires, for the distributive justice reasons described previously, that assistance be targeted at lower emissions development, rather than adaptation, which (again) would only benefit those who receive direct assistance.

The benefits of mitigation are more widely shared than the benefits of adaptation. Those who are worst off are most benefited by it, since they have the least adaptation resources and are often the most exposed to possible ecological harms (for example, natural disasters and inconsistencies in food production). The costs of mitigation can be borne by those who have the greatest economic and technological capacity to respond to anthropogenic climate change, and are also the most responsible for causing it to occur. Prioritizing mitigation over adaptation does not require forestalling economic development to help raise people out of poverty.[5] Therefore, consideration of distributive justice favours mitigation over adaptation.

Human Rights

Distributive justice is not the only (non-welfarist) ethical consideration that favours more intensive mitigation. Simon Caney (2010) has argued that anthropogenic global climate change violates human rights and that this consideration favours mitigation and adaptation over compensation.

A human rights approach requires us to reconceive the way in which one thinks about the costs involved in mitigation and adaptation. Some have argued that it would be extremely expensive to prevent dangerous climate change and hence that humanity should not do this. If, however, it is true that climate change violates

[5] In 2005, 880 million people lived on less than USD 1 Purchasing Power Parity per day (PPP/day), 2.6 billion people on less than USD 2 PPP/day, and 5.15 billion on less than USD 10 PPP/day (World Bank 2008).

human rights then this kind of reasoning is inappropriate. If a person is violating human rights, then he or she should desist even if it is costly. The implications for mitigation and adaptation are clear. That mitigation and adaptation would be *costly* similarly does not in itself entail that they should not be adopted. If emitting greenhouse gases issues is rights violations it should stop, and the fact that it is expensive does not tell against that claim. A human rights approach thus requires us to reframe the issues surrounding the costs of mitigation and adaptation (Caney 2010: 171; see also UN HRC 2009).

In Caney's view, human rights are 'moral thresholds'. Human rights violations are wrongs that cannot be undone, and they are not tradable or substitutable for other goods and values. Since he believes that climate change will result in human rights violations, reducing the magnitude of climate change and its impacts is required so that those violations are minimized. This holds even if aggressive mitigation has high costs and does not maximize overall social welfare.

Whether global climate change will result in human rights violation is more complicated than it might initially seem. The reason for this is that the causal chain from GHG emissions in one place to population displacement (due to sea level rise) in another place, for example, is much more complex, indirect and diffuse than is the case with prototypical human rights violations, for example, genocide, displacement by force, slavery, and rape (Posner and Sunstein 2007). It may be that global climate change undermines human rights, in that it makes it more difficult for some communities to secure food, water, health, housing, and self-determination, without there being a concomitant violation of their rights. As with the moral status of future generations, I will not defend a position here. Instead, I will, again, draw the warranted conditional conclusion: to the extent that global climate change results in human rights violations, mitigation of global climate change is ethically preferable to adaptation to global climate change.

Non-human Species

The background historical rate of extinction is estimated to be approximately one species per million per year (or 0.0001 per cent).[6]

[6] Baillie et al. (2004) calculates the historical rate of extinction as 1-1 E/MSY.

Studies have found that 35 per cent of bird species and 52 per cent of amphibians have traits that put them at increased risk of extinction due to global climate change (Foden et al. 2008); that 20 per cent of lizard species are likely to be extinct by 2080 due to global climate change (Sinervo et al. 2010); and that 15–37 per cent of species will be committed to extinction by 2050 on mid-level warming scenarios (Thomas et al. 2004). Overall, the Intergovernmental Panel on Climate Change (IPCC) concludes that

> there is medium confidence that approximately 20–30 per cent of species assessed so far are likely to be at increased risk of extinction if increases in global average warming exceed 1.5–2.5 C° (relative to 1980–1999) [i.e. lower warming scenarios]. As global average temperature increase exceeds about 3.5 C° [i.e. mid-level warming scenarios], model projections suggest significant extinctions (40–70 per cent of species assessed) around the globe (2007: 54).

Thus, even in optimistic future emissions scenarios, global climate change will dramatically increase non-human species extinctions rates.

If species have a type of value that cannot be substituted or compensated for, and this value is lost through climate change due to species extinctions, then this justifies favouring aggressive mitigation over adaptation, even if it does not maximize overall social utility, in just the same way that consideration of human rights does (Palmer 2011). It is widely recognized that individual species have instrumental value. Some species are medicinally valuable (for example, horseshoe crabs and sweet wormwood); some are economically valuable (for example, honey bees and mahogany); and some are recreationally valuable (for example, rainbow trout and sequoia redwoods). Biodiversity is also instrumentally valuable in myriad ways. For example, plant species richness has been found to enhance eco-system multifunctionality (Maestra et al. 2012), and restoration of biodiversity has been found to increase eco-system services and productivity (Worm et al. 2006; Benayas et al. 2009). However, when something is instrumentally valuable as a means to an end, it is possible to compare it to other potential means to the same end. Moreover, if it is lost, but some other equally adequate means becomes available, then there is no net value loss. Because it is substitutable and compensable, the instrumental value of species does not favour mitigation, adaptation or compensation.

However, it is a common view among environmental ethicists that species also have final value — that is, they are valuable for what they are, and not merely for what they can do for us.[7] In most views, the final value of species is connected to their unique and human independent ecological and evolutionary situatedness (Soulé 1985; Callicott 1989; Elliot 1992; Katz 2000; Rolston 2001). That people value species expressing their distinctive form of life in their evolved ecological context is not speculative. It is evidenced by research on people's environmental values, the resources that people contribute to wilderness protection and *in situ* species conservation, support for species protections that are committed to preserving species in their habitat, and the priority placed on *in situ* preservation in conservation biology (Bosso 2005). An implication of the value of species being tied to their ecological and evolutionary situatedness is that they retain their value only so long as those relationships remain intact and are not disrupted by human activities. As a result, it is not preservation of *species* that we want, but the preservation of *species in the system*. It is not merely *what* they are, but *where* they are that humans must value correctly. The species can only be preserved *in situ*; the species *ought* to be preserved *in situ* (Rolston 2001: 411).

Even if a price can be put on something that has final value — that is, there is some finite value people are willing to pay to buy it or that they would take to sell it — the price does not fully capture how or why it is valued. So it is not the case that the final value of species that are lost as a result of anthropogenic climate change can be substituted for or justified by other values later, such as benefits to others or costs avoided. Moreover, because the final value of species is based on their ecological and evolutionary properties, it cannot be replicated elsewhere. Therefore, the final value of species favours responses to climate change that forestall these in-site relationships being broken, that is, mitigation over adaptation. The strength of this justification for prioritizing mitigation is dependent upon the scope and magnitude of the final value of species and biodiversity. However, to the extent that they have final value, and to the extent that the value is connected to their evolved ecological

---

[7] It is common in environmental ethics to refer to non-instrumental value as 'intrinsic value'. However, for reasons I discuss at length elsewhere, the intrinsic value terminology is ambiguous and misleading (Sandler 2012).

relationships, the value of non-human species favours mitigation over adaptation.

The overall case made here for mitigation over adaptation is largely conditional. To the extent that (*a*) the welfare of future generations, (*b*) distributive justice, (*c*) human rights, and/or (*d*) the value of non-human species should be taken into consideration when determining policy responses to global climate change, mitigation should be favoured over adaptation. However, as indicated in these discussions, the ethical commitments are both reasonable and widespread. Moreover, denying all of them results in an impoverished view of what matters ethically — that is, only the welfare of present human beings, without consideration of distribution of benefits and harms.

How to Mitigate

Not all mitigation strategies are equal. One reason for this is technical. Some mitigation strategies are less costly, and more immediately and easily implementable, scalable, and/or fecund (that is, conducive to additional emissions reductions in the future) than are others.[8] For example, increasing vehicle and building efficiency can be accomplished with already available technologies, and solar energy production is highly scalable and fecund, due to the amount of solar radiation available and the fact that its widespread deployment involves developing technologies and infrastructures that will facilitate further use. However, technical considerations are not the only ones relevant to evaluating possible emissions reduction strategies and policies. Not all mitigation strategies are ethically equal. In this section, I briefly outline considerations that are relevant to the ethical evaluation of possible mitigation strategies and policies.

Ancillary Benefits

Approaches to emissions reductions that have ancillary social and ecological benefits — that is, benefits in addition to GHG mitigation — are preferable to those that do not. For example, the United

---

[8] Scalability and fecundity are particularly crucial, since the largest reductions in emissions from BAU must occur in the second half of the century. Technologies and infrastructures that are developed and put into place prior to 2050 must establish an emissions reduction pathway that can accelerate thereafter (Davis et al. 2010).

Nations' REDD programme (United Nations Reducing Emissions from Deforestation and Forest Degradation in Developing Countries) aims to reduce GHG emissions by decreasing the amount of deforestation from what would occur on BAU (approximately 17 per cent of global emissions are attributable to deforestation) (IPCC 2007). However, preventing deforestation is crucial to preserving biodiversity and eco-system services. It also helps indigenous people to remain on their lands, and can increase social and ecological systems' adaptive capacity (UN REDD 2009). Moreover, REDD involves resource transfers from developed to developing countries in exchange for deforestation reductions. In this way, it is another mechanism for encouraging lower emissions development.

Another approach to mitigation that has significant ancillary benefits is decreasing population growth — through increasing women's access to family planning, education, and opportunities in the work force and civic life. Population growth is one of the primary drivers of growing emissions on BAU, particularly with respect to emissions in less developed nations (IPCC 2007). Therefore, reducing population growth rates from BAU has a very large mitigation potential (Cafaro 2011). However, not all policies for reducing population growth are ethically equal. Some possible policies are coercive and harmful, for example, stringent one child per family policies and sterilization programmes. Other policies, such as those that increase access to medical care, family planning, education, and civic life for women, are not only strongly correlated with reduction in birth rates, but also empower women, increase their autonomy and improve their health (Carter 2004; Cafaro 2011).

Agriculture is another potential source for significant emissions reductions, since it is responsible for approximately 14 per cent of global CO_2 emissions and 18 per cent of CO_2e emissions (UN FAO 2006; IPCC 2007). One way to reduce agricultural emissions is to decrease meat consumption from the BAU scenario, particularly for meat produced through intensive animal feed operations. The reason for this is that getting nutrition through intensively farmed meat is highly inefficient, and 80 per cent of agricultural emissions are attributable to animal agriculture (UN FAO 2006). Agricultural animals use feed inputs for all sorts of metabolic activities in addition to tissue growth. As a result, it is several times more efficient to consume plant protein, calories and nutrition directly, rather than pass it through an agricultural animal (Lappé 1982; World Watch Institute 2004). Consuming less meat could, therefore, decrease emissions associated with both crop and animal agriculture, for example, methane emissions from livestock (Cafaro 2011). Moreover, the reduction in crop and animal agriculture would have significant ancillary benefits, since there are a number of problematic effects associated with industrial crop and animal agriculture, for example, habitat destruction, nitrogen eutrophication, waste runoff (from pesticides, herbicides and animals), and animal suffering.

The foregoing does not constitute full ethical evaluations of the mitigations strategies and policies discussed. However, it illustrates the ways in which some mitigation efforts have robust ancillary social and ecological benefits. To the extent that a strategy has such benefits — that is, protects or promotes autonomy, justice, flourishing (human and non-human), public health, and other values, it is preferable to those that do not.

Unintended Negative Impacts

Some mitigation strategies are much more likely to have negative unintended impacts than others. In some cases, negative ecological impacts are foreseeable. For example, past experience with hydroelectric power makes it clear that increasing energy supply from hydroelectric dams will have severe effects on water levels and flows in riparian eco-systems, with detrimental impacts on ecological and human communities. In other cases, negative ecological effects may be likely, but difficult to predict. For example, geoengineering approaches to mitigation that involve intensive and large scale intervention into complex ecological and climatic processes, such as iron ocean fertilization and solar radiation management, are likely to have more negative ecological impacts than less control-oriented and interventionist approaches to mitigation, for example, eating less meat and increasing vehicle efficiency. It may be difficult to predict in advance what the problematic effects will be, but the nature of the intervention opens it to greater risks than other types of strategies.

Policies to foster slower (or negative) global economic growth would significantly reduce GHG emissions from BAU. However, they would also have substantial and predictable detrimental social impacts on the global poor. For this reason, mitigation policies should be focused on promoting economic development along lower emissions pathways (which has ancillary benefits), or transferring resources (for example, educational and technological) from those with excess to those in poverty (or shifting expenditures away from the trivial and detrimental, such as military spending, to more developmental spending). Similarly, a carbon tax policy that would aim to reduce consumption by increasing the cost of goods and services could be detrimental to many people living in more economically developed nations, which also have large populations living in poverty.[9] Therefore, a carbon tax approach to emissions reductions would need to include an exemption for basic goods, or else provide individual emissions allocations that are sufficient for basic goods acquisition.

Again, the forgoing is not an all-things-considered evaluation of the GHG reduction strategies and policies discussed, but rather illustrates the ways in which some mitigation approaches can have detrimental social and ecological effects.

Coerciveness, Constraint, and Demandingness

As the discussion of possible population growth reduction policies indicates, some policies that could realize significant mitigation increase people's choices, opportunities, and autonomy. However, other possible approaches to mitigation would constrain people's choices. For example, one way to reduce emissions is to prevent the increase in air travel that is expected on BAU. Air travel accounts for only 3 per cent of global CO_2 emissions, but is among the fastest growing sources of emissions (IPCC 2007). One way to reduce this increase would be to put a hard cap on the number of non-essential flights that a person could take each year (Cafaro 2011). However, this restricts global mobility and individual choice about resource use. An alternative, which is already being implemented as part of the European Union's (EU) Emissions Trading System (ETS), is to add a carbon emissions surcharge to flights, so that their ecological costs are internalized and passed on to those who are responsible for travelling. This would have the effect of increasing the cost of air travel, which will affect travellers (or potential travellers) with

---

[9] The poverty rate in the United States in 2010 was 15.1 per cent, for example (US Census Bureau 2011).

lower incomes more than those with higher incomes. However, it may reduce air travel, and the surcharge could be used for developing aviation technologies (for example, fuels) that would reduce emissions, or for offsetting emissions generally (as is being done with the EU ETS surcharge).

Related to the issue of coerciveness is that of demandingness. All other things being equal, mitigation strategies that are less demanding on individuals or organizations are preferable to those that require them to make significant efforts and sacrifices. This consideration will tend to favour technologically-oriented mitigation. For example, the allure of geoengineering and inexpensive and abundant zero emission energy sources is that they promise to address the problem of global climate change without requiring sacrifices by individuals or alterations of social or economic systems.

In general, policies that involve hard prohibitions and limits on behaviour are less preferable than those that involve incentivizing, and policies that increase people's choices and opportunities are more preferable to those that decrease their autonomy and are more demanding.

Conclusion

In policy discussions regarding mitigation, the focus has been on how much to mitigate, as well as how to fairly distribute mitigation responsibilities between nations. These are crucial issues, and in the first part of this chapter I argued that several ethical considerations, including the welfare of future generations, distributive justice, human rights, and the value of non-human species, favour aggressive mitigation over adaptation. However, even after mitigation goals are set and responsibilities allocated, there is the further issue of how best to pursue mitigation reductions. Some considerations relevant to mitigation evaluation are technical — for example, cost, feasibility, ease and immediacy of implementation, magnitude (or scalability), and fecundity. However, other considerations are ethical. Policies that have ancillary social and ecological benefits are preferable to those that do not; less constraining policies are preferable to those that are more coercive; and strategies that are less disruptive of ecological and social systems are preferable to those with higher ecological and social risks.

References

Baer, Paul, Tom Athanasiou, Sivan Kartha, and Eric Kemp-Benedict. 2008. *The Greenhouse Development Rights Framework: The Right to Development in a Climate Constrained World*, Heinrich Böll Foundation, Christian Aid, EcoEquity and the Stockholm Environment Institute, Berlin.

Baillie, Jonathan E. M., Leon A. Bennun, Thomas M. Brooks, Stuart H. M. Butchart et al. 2004. *A Global Species Assessment*. International Union for Conservation of Nature (IUCN), UK.

Benayas, José M. R., Adrian C. Newton, Anita Diaz, and James M. Bullock. 2009. 'Enhancement of Biodiversity and Ecosystem Services by Ecological Restoration: A Meta-Analysis', *Science* 325(5944): 1121–24.

Bosso, Christopher J. 2005. *Environment Inc.: From Grassroots to Beltway*, University Press of Kansas, Kansas.

Cafaro, Philip J. 2011. 'Beyond Business as Usual: Alternative Wedges to Avoid Catastrophic Climate Change and Create Sustainable Societies', in Denis G. Arnold (ed.), *The Ethics of Global Climate Change*, Cambridge University Press, Cambridge, 192–215.

Callicott, J. Baird. 1989. *In Defense of the Land Ethic: Essays in Environmental Philosophy*, State University of New York Press, Albany.

Caney, Simon. 2010. 'Climate Change, Human Rights, and Moral Thresholds', in Stephen Gardiner, Simon Caney, Dale Jamieson, and Henry Shue (eds), *Climate Ethics*, Oxford University Press, Oxford, 163–77.

Carter, Alan. 2004. 'Saving Nature and Feeding People', *Environmental Ethics* 26(4): 339–60.

Climate Interactive. 2011a. 'Possibilities for the Global Climate Deal', http://climateinteractive.org/scoreboard/scoreboard-science-and-data/graphs-possibilities-for-the-global-climate-deal (accessed 5 January 2012).

———. 2011b. 'Data and References: December 9th, 2012 Release', http://www.climateinteractive.org/scoreboard/scoreboard-science-and-data/data-and-references (accessed 9 February 2012).

Costa, Luìs, Diego Rybski and Jürgen P. Kropp. 2011. 'A Human Development Framework for CO2 Reductions', *PLoS ONE* 6(12): 1–14.

Davis, Steven J., Ken Caldeira and H. Damon Matthews. 2010. 'Future CO_2 Emissions and Climate Change from Existing Energy Infrastructure', *Science* 329(5997): 1330–33.

Elliot, Robert. 1992. 'Intrinsic Value, Environmental Obligation, and Naturalness', *The Monist* 75(2): 138–60.

Foden, Wendy B., Georgina M. Mace, Jean-Christophe Vié, Ariadne Angulo et al. 2008. 'Species Susceptibility to Climate Change Impacts' in Jean-Christoph Vié, Craig Hilton-Taylor and Simon N. Stuart (eds), *Wildlife in a Changing World: An Analysis of the 2008 IUCN Red List of Threatened Species*, IUCN, Gland, 77–88, http://data.iucn.org/dbtw-wpd/edocs/RL-2009-001.pdf (accessed 15 March 2011).

Gillett, Nathan P., Vivek K. Arora, Kirsten Zickfeld, Shawn J. Marshall, and William J. Merryfield. 2011. 'Ongoing Climate Change Following a Complete Cessation of Carbon Dioxide', *Nature Geoscience* 4: 83–87.

Hansen, James, Reto Ruedy, Makiko Sato, and Ken Lo. 2010. 'Global Surface Temperature Change', *Reviews of Geophysics* 48(4): 1–29.

Intergovernmental Panel on Climate Change (IPCC). 2007. *Climate Change 2007: Synthesis Report*, IPCC, Geneva.

Joughin, Ian and Rober B. Alley. 2011. 'Stability of the West Antarctic Ice Sheet in a Warming World', *Nature Geoscience* 4: 506–13.

Katz, Eric. 2000. 'The Big Lie', in William Throop (ed.), *Environmental Restoration: Ethics, Theory, and Practice*, Humanity, Amherst, 83–93.

Kempton, Willet, James S. Boster and Jennifer A. Hartley. 1995. *Environmental Values in American Culture*, MIT Press, Cambridge.

Lappé, Frances M. 1982. *Diet for a Small Planet*, Ballantine Books, New York.

Lomborg, Bjørn. 2007. *Cool It: The Skeptical Environmentalist's Guide to Global Warming*, Knopf Publishing Group, New York.

———. 2001. *The Skeptical Environmentalist: Measuring the Real State of the World*, Cambridge University Press, Cambridge.

Maestra, Fernando T., José L. Quero, Nicholas J. Gotelli, Adriàn Escudero et al. 2012. 'Plant Species Richness and Ecosystem Multifunctionality in Global Drylands', *Science* 335(6065). 214–18.

National Oceanic and Atmospheric Administration (NOAA). 2011. 'NOAA's 1981–2010 Climate Normals', http://www.ncdc.noaa.gov/oa/climate/normals/usnormals.html (accessed 27 April 2012).

Nordhaus, William. 2007a. 'Critical Assumptions in the Stern Review on Climate Change', *Science* 317(5835): 201–2.

———. 2007b. 'The Stern Review on the Economics of Climate Change', *Journal of Economic Literature* 45(3): 686–702.

Pacala, Stephen W. and Robert H. Socolow. 2004. 'Stabilization Wedges: Solving the Climate Problem for the Next 50 Years with Current Technologies', *Science* 305(5686): 968–72.

Palmer, Clare. 2011. 'Does Nature Matter? The Place of the Non-human in the Ethics of Climate Change', in Denis G. Arnold (ed.). *The Ethics of Global Climate Change*, Cambridge University Press, Cambridge, 271–91.

Posner, Eric A. and Cass R. Sunstein. 2007. 'Climate Change Justice', *The Georgetown Law Journal* 96(5): 1565–1612.

Pritchard, Hamish D., Robert J. Arthern, David G. Vaughan, and Laura A. Edwards. 2009. 'Extensive Dynamic Thinning on the Margins of the Greenland and Antarctic Ice Sheets', *Nature* 461(7266): 971–75.

Rolston III, Holmes. 2001. 'Biodiversity', in Dale Jamieson (ed.), *A Companion to Environmental Philosophy*, Blackwell Publishers, Oxford, 402–15.

Romm, Joe. 2008. *The Self-Limiting Future of Nuclear Power*, Center for American Progress, Washington, http://www.americanprogressaction.org/issues/2008/pdf/nuclear_report.pdf (accessed 17 February 2012).

Sandler, Ronald. 2012. *The Ethics of Species*, Cambridge University Press, Cambridge.

Sinervo, Barry, Fausto Mendez-de-la-Cruz, Donald B. Miles, Benoit Heulin et al. 2010. 'Erosion of Lizard Diversity by Climate Change and Altered Thermal Niches', *Science* 328(5980): 894–99.

Soulé, Michael E. 1985. 'What is Conservation Biology?', *Bioscience* 35(11): 727–34.

Stern, Nicholas. 2007. *Stern Review: The Economics of Climate Change*, Cambridge University Press, Cambridge, http://webarchive.nationalarchives.gov.uk/+/http://www.hm-treasury.gov.uk/d/Chapter_9_Identifying_the_Costs_of_Mitigation.pdf (accessed 28 January 2014).

Thomas, Chris D., Alison Cameron, Rhys E. Green, Michael Bakkenes, et al. 2004. 'Extinction Risk from Climate Change', *Nature* 427(6970): 145–48.

United Nations Food and Agriculture Organization (UN FAO). 2006. *Livestock's Long Shadow: Environmental Issues and Options*, http://www.fao.org/docrep/010/a0701e/a0701e00.HTM (accessed 17 February 2012).

United Nations Framework Convention on Climate Change (UNFCCC). 2012a. 'Transitional Committee for the design of the Green Climate Fund', http://unfccc.int/cooperation_and_support/financial_mechanism/green_climate_fund/items/5869.php (accessed 17 February 2012).

———. 2012b. 'Clean Development Mechanism', http://cdm.unfccc.int/index.html (accessed 9 February 2012).

———. 2009. 'Copenhagen Accord', http://unfccc.int/resource/docs/2009/cop15/eng/l07.pdf (accessed 9 February 2012).

United Nations Human Rights Council (UN HRC). 2009. 'Human Rights and Climate Change', http://ap.ohchr.org/documents/E/HRC/resolutions/A_HRC_RES_10_4.pdf (accessed 9 February 2012).

United Nations Reducing Emissions from Deforestation and Forest Degradation in Developing Countries (UN REDD). 2009, http://www.un-redd.org/aboutredd/tabid/582/default.aspx (accessed 27 April 2013).

United States (US) Census Bureau. 2011. 'Poverty', http://www.census.gov/hhes/www/poverty/about/overview/index.html (accessed 17 February 2012).

United States Environmental Protection Agency (US EPA). 2010. 'EPA Analysis of the American Power Act in the 111th Congress', http://www.epa.gov/climatechange/Downloads/EPAactivities/EPA_APA_Analysis_6-14-10.pdf (accessed 27 April 2013).

World Bank. 2008. *World Bank Development Indicators*, http://books.google.com/books?id=O67oDJW01pwC&printsec=frontcover&source=gbs_ge_summary_r&cad=0#v=onepage&q&f=false (accessed 26 January 2012).

World Watch Institute. 2004. *State of the World 2004: Special Focus, The Consumer Society*, Norton, New York.

Worm, Boris, Edward B. Barbier, Nicola Beaumont, J. Emmett Duffy et al. 2006. 'Impacts of Biodiversity Loss on Ocean Ecosystem Services', *Science* 314(5800): 787–90.

5

Responsibility for Mitigation and Adaptation, and the Right to Sustainable Development

Darrel Moellendorf

There is a fundamental tension in international negotiations around a comprehensive climate change treaty. Developing and least developed countries look to industrialized states both to take the lead in addressing the problem and to assume the bulk of the burdens of doing so. But several such countries refuse to be party to an agreement that does not contain legal commitments for developing and least developed countries. And leading developing countries eschew any treaty that places legal requirements on them to constrain their emissions. This tension led to disappointment at the Copenhagen Conference of the Parties (COP 15) of the United Nations Framework Convention on Climate Change (UNFCCC) in 2009,[1] where there had been hopes of negotiating a comprehensive treaty to take effect after the first commitment period of the Kyoto Protocol. More recently, at COP 17 in Durban an agreement that would bring all parties into a legal document was decided upon only by postponing both discussion of its content to future meetings and the date at which it should go into effect by several years.

Important matters of political morality are at stake in these diplomatic wranglings. One of the most important concerns is how to assign responsibility for mitigating and financing adaptation to climate change. After examining the challenges, I shall contend that accounts of responsibility that are tied to the historic emissions of states are morally inadequate. Instead, I shall defend the application of the ability-to-pay principle. Finally, I shall argue that this

[1] I use 'UNFCCC' to refer to the United Framework Convention as an institution and 'Convention' to refer to the United Nations Framework Convention treaty document.

conception of responsibility fits nicely with the right to sustainable development, which would be unreasonable for parties to the UNFCCC process to ignore.

The Challenges of Mitigation and Adaptation

Millions of people are already at risk of extreme weather and flooding. Currently around 344 million people are exposed to tropical cyclones, 521 million to floods, 130 million to droughts, and 2.3 million to landslides (UNDP 2007: 98). Climate change is expected to increase these numbers very significantly. Hundreds of millions of people are at risk of inundation from tropical storms and rising sea levels. The devastation caused by drought, flooding, sea-level rise, tropical storms, and changed disease vectors could result in long-term setbacks to human development in many poor societies (ibid.: 88–89).

Any effective international treaty for climate change mitigation will have to lower global CO_2 emissions very dramatically. To do this, the cost of fossil fuels relative to renewables will have to increase. But human development is immensely energy intensive. A climate change treaty that significantly raises the costs of energy in the developing world could also hinder or stall human development.

In both the Copenhagen Accord (noted at the COP 15) and the documents that were agreed upon at the COP 16 in Cancun in 2010, the primary goal of climate change mitigation is expressed as limiting anthropogenic warming to 2° Celsius above pre-industrial levels. To keep concentrations within the range identified by the Intergovernmental Panel on Climate Change (IPCC) would require steep reductions in total global emissions, starting in the next couple of years, with the goal of emission levels in 2050 at 50 to 85 per cent below 2000 levels (IPCC 2007: 20). There is good reason to believe that such reductions are technically feasible through a combination of conservation, changed land use, and alternative energy generation (Carbon Mitigation Initiative). Moreover, reputable economic analyses forecast that total global economic costs would be very modest. The *Stern Review*, an economic study conducted in the United Kingdom, and endorsed by many independent economists, forecasts only very modest costs associated with these reductions. The average of various costs scenarios projects world economic growth to be only 1 per cent lower per annum than it would be absent the investments necessary to mitigate climate change (Stern 2007: 211).

In addition to the costs of mitigation, resulting from changing land and energy use, there will be costs of adaptation, resulting from climate change caused damage and measures taken to avoid the damage, respectively. Adaptation costs of avoiding the damage must be taken up soon while the adaptation costs of responding to damage will continue indefinitely. Identifying the latter poses severe epistemic problems, since in its present state climate science does not issue in fine-grained analyses of the consequences of mean global temperature increases. But the best scientific forecasts suggest significant social disruptions, even given just a two degree temperature rise: hundreds of millions of people exposed to increased water stress; decreased crop yields in low latitudes; millions more people exposed to coastal flooding; increased morbidity and mortality from heat waves; and a heavier burden from malnutrition, diarrheal, cardio-respiratory and infectious diseases (IPCC 2007: 10). In other words, the adaptation costs of responding to climate change will be heavy, even if we manage to hit an ambitious mitigation goal, if we do not also take early measures to invest in adaptation to avoid some of these outcomes.

Conceptions of Responsibility

Climate change mitigation must be a collective international effort and its costs must somehow be distributed internationally. Climate change adaptation will also be expensive. In the absence of an international commitment to fund adaptation, these costs will be born mostly by the states in which the suffering is the greatest. The distribution of both mitigation and adaptation costs could be affected by policy decisions. The possibility of changing the distribution of costs raises the moral question of who should bear responsibility for them.

One principle of responsibility familiar from other environmental policy discussions is the 'polluter-pays-principle'. This principle might be defended on either fault or no-fault grounds. Fault conceptions of responsibility are common in the law of torts, which seeks to assign responsibility for accidents against a stable background of property entitlements. Holding an agent to be at fault for a circumstance usually has, at least, three necessary conditions:

(*a*) causation — the agent brought about the circumstance as a consequence of her actions; (*b*) voluntariness — the agent did so voluntarily; and (*c*) knowledge — the agent knew (or should have known) the consequences of his/her action.[2] Defenders of fault conceptions of responsibility sometimes claim that fault must be present if a sanction against an agent is not to be a violation of her liberty. Simply put, liberty requires that a person should not be sanctioned for actions unless she is at fault. Appeals to efficiency provide a different kind of justification of fault conceptions of responsibility. By assigning responsibility only to those who voluntarily and knowingly create problems, we establish a system of incentives that reduces the incidence of such misdoing and we lower the incidence of externalities — the costs of misdeeds being passed along to others.

Despite the prevalence of fault-based conceptions of responsibility in the political discourse about climate change, there are two significant problems with recouping the costs of climate change by means of assigning responsibility on the basis of fault (see Caney 2005). The first has to do with satisfying the causation requirement. People living now did not cause all, perhaps not even most, of the problem. The average residence of a CO_2 molecule in the atmosphere is a hundred years or more. Hence, much of the damage-causing stock of atmospheric CO_2 was produced by people who are now long dead. We can only recoup costs from individuals who are alive; and the living caused only a portion of the damage.

A second problem arises in relation to satisfying the knowledge condition. Among the still-living, many fail to meet the knowledge condition for some of their early emissions. General knowledge that climate change is produced by greenhouse gases is relatively recent. In 1988, the United Nations General Assembly adopted Resolution 43/53:

> Noting with concern that the emerging evidence indicates that continued growth in atmospheric concentrations of 'greenhouse' gases could produce global warming with an eventual rise in sea levels, the effects of which could be disastrous for mankind if timely steps are not taken at all levels (ibid.).

[2] For more on the application of the distinction between fault and no-fault accounts in climate change, see Shue (1993).

That same year the IPCC was founded by the World Meteorological Association and the United Nations Environmental Programme. The first assessment report of the IPCC was not issued until 1990. At most, a fault account of responsibility would seem to be limited to the emissions of those people currently living, and only their emissions, since the late 1980s or early 1990s. This leaves responsibility for the costs of earlier emissions unassigned. A fault conception of responsibility applied to individual persons simply cannot recoup costs for all of the damages caused by climate change.

The first of the two problems discussed earlier can be at least partially avoided if the bearers of fault are not individual people (now long dead) but states.[3] Causation for CO_2 emissions that occurred prior to the lifetime of any living person could be attributed to those states that have had a continuous existence since the industrial revolution. This would also be consistent with the state-centred language of international treaties. Nonetheless, there would still be two limitations preventing recoupment of all historic costs. First, the assignment of responsibility to a particular state could go back no further than its origin. And, not all historic emissions originated from currently existing states. Second, states — or their leaders — also fail to meet the knowledge condition before the late 1980s and early 1990s.

In addition to these two limitations, there is significant moral tension between the practice of holding states (rather than individuals) at fault and the rationales for using a fault conception. It is nearly impossible to assign responsibility to states without preventing the costs from devolving eventually to the citizens of the states since state revenues are raised by taxing citizens. In practice, then, the responsibility devolves with the costs. So, were an international institutional arrangement to lay fault on a state for all its emissions since, say, 1988, it would, for practical purposes, assign costs to the citizens of that state, an increasingly large proportion of which were born after that date. If the fault conception of responsibility is appropriate because by sanctioning individuals only if they are at fault their liberty is respected, then the practice of devolving the costs of emissions to individuals is undermining of the liberty-

[3] See Vanderheiden (2008: ch. 5) for a defence of a fault conception of collective responsibility.

respecting justification of the conception. Alternatively, if the fault conception of responsibility is justified because by holding agents responsible for the harms they cause we create incentives not to cause harms, a problem still arises for the practice of holding states at fault. No such incentives are created by devolving costs to individuals who did not cause the harm; but that is just what is done when costs devolve to citizens who were not alive when the emissions occurred.

In light of the limitations and problems of a fault conception of responsibility, a no-fault account of responsibility for historic emissions might seem more promising. No-fault conceptions of responsibility vary, but they have in common a denial of at least one of the three conditions (stated earlier) for assigning responsibility on the basis of fault. One such conception employs strict liability, which holds agents responsible if they cause a problem, regardless of whether they acted with knowledge.[4] Given the problems with the knowledge condition, strict liability might seem to be a better basis for the polluter-pays-principle. One method is to apply strict liability to states, holding them responsible for historic emissions in proportion to their percentage of contribution to atmospheric stocks of greenhouse gases.

The general challenge to the justification of strict liability conceptions of responsibility is to account for why bare causation is enough to hold an agent responsible for what she has done, especially if he/she is not at fault. One response to the challenge relies on the special risks associated with the activity and the existence of prior notice. In the law, strict liability is usually applied only with respect to activities that are especially dangerous or important to human health, in which a high standard of care is warranted, and when agents can be put on notice before hand that they will be held responsible for the negative effects of their actions (even if they are not at fault). Prior notice goes some way towards alleviating concerns that responsibility based on causation alone is unfair. The idea is that those engaging in such activity are knowingly assuming the risks of being held responsible. This defence does not work in the case of emissions prior to general knowledge of their danger, since agents could not have been be put on notice that they would

[4] For more on strict liability and responsibility for climate change, see Weisbach (2010).

be held responsible for the unknown dangers. In the absence of such notice, strict liability suffers from being a form of *ex post facto* sanction.

Beneficiary responsibility is another no-fault conception of responsibility that maintains a connection with historical emissions. Beneficiary responsibility jettisons the requirement of causation but still maintains a connection to past action through the requirement that those to be held responsible must have benefited from the particular action.[5] Perhaps, persons living in industrialized countries can be held responsible for past greenhouse gas emissions because they benefit from the standards of living produced by past emissions. A justification of this conception of responsibility requires a compelling reason to believe that a connection with a past action, even as tenuous as that one has benefited from it, should be the grounds for assigning one's responsibility for it. For one might benefit from action that one had no control over and which one might not have supported if one had a say. Suppose I benefit from my neighbour painting his house a colour I dislike because keeping up his house serves to maintain the property value of my house. It seems implausible that I am responsible for paying for a portion of his painting costs. It is difficult to see how mere benefit establishes grounds for holding the beneficiary responsible. The case for beneficiary responsibility may seem more compelling when there is great inequality between the beneficiary and others. But to the extent that such is the case, the justificatory work seems to be done by the inequality rather than the benefit.

Fault, strict liability and beneficiary responsibility are three ways to hold agents responsible for historic emissions and thereby buttress something like the polluter-pays-principle for greenhouse gas emissions. Each of them contains problems that a fully adequate account would have to overcome. Setting those matters aside, the polluter-pays-principle is subject to the following important general problem. The principle can direct the assignment of the burdens of reducing CO_2 emissions (or financing adaptation) but it is silent on the permission to emit CO_2 in order to fuel poverty eradicating economic growth (Moellendorf 2009a). Various conceptions of global justice that aim either to eradicate severe poverty or to reduce inequalities between people around the world

5 For a use of beneficiary responsibility, see Neumayer (2000: 189).

support laying responsibility for emissions reductions more heavily on industrialized countries in order to allow developing and least developed countries the leeway to pursue macro-economic policies that promote economic growth and poverty eradication, even if it is by increasing emissions (Moellendorf 2009b). The polluter-pays-principle appears insensitive to this concern of global justice.

In contrast to the polluter-pays-principle, the ability-to-pay principle has the virtue of fitting well with conceptions of global justice that condemn severe poverty and massive inequality. It assigns responsibility in proportion to an agent's capacity (variously measured). The ability-to-pay conception is often used in the assignment of financing state activities, such as defence against various threats and providing for certain aspects of the well-being of citizens. Generally, in financing programmes directed to meet these aims, states do not look for citizens who are at fault for the wealth that they possess. Instead, progressive income taxation is typically defended on grounds that the wealthier have a greater ability to pay.

These comments suggest a second reason in favour of ability-to-pay, namely that it fits particularly well with the project of reforming the basic structure of international relations. Fault and strict liability conceptions of responsibility fit best in the law of accidents in which a stable background of entitlements to property is supposed. They do not fit well in determining what that background should be, in establishing entitlements under a reformed basic structure. Pollution within states may be handled reasonably well by the law of accidents. But determining access to energy and entitlements to use of various forms energy among states, when the well-being of billions of people is at stake, more closely resembles establishing or reforming the background basic structure of international relations between states, than it does deciding how to hold agents responsible for accidents.

A third reason in favour of this principle, in the context of an international climate change regime, is that there is a basis for it in the Convention, which recognizes the 'respective capabilities' of various parties (art. 3.1) and asserts 'the right to sustainable development' (art. 3–4). The first suggests something like an ability-to-pay conception of responsibility for mitigation and the financing of adaption to climate change. The second seems to require that such institutions be consistent with macro-economic policies in developing and least developed countries that seek poverty-eradicating growth.

There is still the matter of whether responsibility on the basis of the ability-to-pay should be assigned to states or to individuals. Here, there are some competing considerations of morality and practicality. Insofar as we prize the well-being or dignity of individuals, we find the moral ideals of equality amongst individuals and poverty eradication for persons attractive. However, an international climate change regime will be the product of treaty negotiation between states that are quite likely to jealously guard their traditional sovereign powers. Moreover, it might be difficult to work out the international institutions for holding individuals responsible, even if states were willing. This makes it doubtful that the assignment of responsibility for the costs of climate change through international negotiation can effectively lay burdens directly on individuals.

This consideration is not merely a concession to realism at the expense of justice. The problems that need solving are practical. To some extent, then, a conception of responsibility also has to be workable. The moral merits of a conception of responsibility would be diminished by its lack of practicality. Additionally, the state-centric language of the Convention concerning the respective capabilities of states and the right to sustainable development seems to provide protections for the well-being of the populations of states; to the extent that such language forms the basis of an assignment of basic responsibility to states, the assignment is sensitive to, even if not directly laid on individuals.

One objection to the assignment of responsibility to states rather than to individuals is, as we have seen, the charge that responsibility would unfairly devolve to individuals. A consequence of assigning responsibility to states is that wealthy people in poor states will carry a less heavy burden than wealthy people in wealthy states and that poor people in wealthy states will carry heavier burden than poor people in poor states. Given a broadly individualist background moral account, the differential assignment of burdens to individuals on the basis of state of residence is worrisome. There are, I think, two considerations that dampen this worry. The first is that it remains a question of state policy how the burdens get assigned to individuals. It is not a necessary feature of this proposal that wealthy people in poor states are given a free ride or that poor people in rich states are burdened. Insofar as responsibility is assigned to states, political processes of the state must take over the assignment of the burdens to their citizenry.

The second consideration is that negotiating a comprehensive climate change treaty under current geo-political conditions, and for the foreseeable future, is a task of international co-ordination. Effective mitigation responses will happen (if they happen) in the next five to 10 years. There is no reason to suppose that during that time span the international state system will be replaced by something better representing the interests of all the individuals of the world. It makes moral sense in this context to employ conceptions of responsibility directed to states. If the result of doing that is an institutional order that results in differential burdens assigned to persons on the basis of a morally arbitrary feature such as state residence, then that would not be peculiar to the ability-to-pay conception of responsibility. Rather, it applies to all proposals directed towards the only agents — states — capable of dealing with the problem of climate change.

The Right to Sustainable Development

As I noted in the first section, a principal reason for an international regime that mitigates climate change is the threat that climate change poses to human development. But human development is threatened both by the devastation that climate change might wreak and by the international policies that would raise the costs of consuming energy. Progress in achieving a high level of human development is tremendously energy intensive. In order to develop rapidly, countries need access to cheap forms of energy. Currently, fossil fuels are much cheaper than renewable energy. In order that the human development process does not slow down, energy prices for developing and least developed countries must not increase significantly in the near term.

There are two ways that an international treaty mitigating climate change could facilitate the consumption of energy at low prices in such countries. The price of fossil fuels for these countries could be kept down by not decreasing supply through consumption restrictions. Or, the production and consumption of more expensive renewable energy could be subsidized. There seems to be a third possibility also: there could be a technological breakthrough that allows for the production of renewable energy at prices that are comparable to fossil fuel. But that is unlikely to come about in the absence of policy measures that raise the price of fossil fuels relative to forms of renewable energy in the industrialized world, since such measures would encourage more investment in research and development into renewable energy. Thus, the apparent third possibility is not a genuine rival to the other two means of keeping renewable energy prices low in the developing world, but a possible fortuitous consequence of them.

Both methods (in the two treaties) for not raising energy-related costs involve assigning responsibility for climate change mitigation largely to advanced industrialized or most highly human-developed societies. Keeping the price of fossils fuels down in developing countries and not restricting consumption (or not restricting it much) would permit growth in emissions in the developing world. In that case, the only way a 50 to 85 per cent global reduction could be obtained would be through very deep reductions, quite possibly in excess of 100 per cent, in highly developed countries.[6] Alternatively, keeping the price of renewable energy down in developing countries would require subsidies in the form of resource, technology or intellectual property transfers from developed to developing and least developed countries. Either of these methods would be consistent with the Convention's affirmation of the right to sustainable development, insofar as sustainability would be obtained through an international mitigation regime and development would be permitted by assigning responsibility for the international regime in accordance with a conception of the ability-to-pay principle, where ability varies with level of human development.

There is controversy surrounding the view that an international climate change treaty should seek both to mitigate climate change and safeguard the possibility of human development. Eric Posner and David Weisbach (2010) argue that the only appropriate goal of an international climate change agreement is efficient mitigation. They understand efficiency in Pareto terms applied to states; at least one state should be made better off as a result of the climate change agreement and no state should be made worse off (ibid.). They take the measurement of better and worse to be in comparison to likely future gross domestic product (GDP) under

[6] For one proposal that foresees such reductions, see Baer et al. (2008).

business-as-usual (energy policy without additional mitigation) growth. This goal would be permissive of an international agreement that allowed policy-induced serious setbacks to human development in some states, just so long as those states would be even worse off in the absence of an agreement. Since geography and poverty conspire to make many very poor people in deeply impoverished countries especially vulnerable to terrible suffering under the business-as-usual scenario, the Posner–Weisbach proposal would tolerate either changes to climate that amount to only minor improvement over that suffering or a climate change treaty that raised the price of energy for developing and least developed countries such that human development was only slightly better than what one would obtain given the ravages of climate change. Seeking to secure the possibility of human development, according to this view, is an effort at redistribution that distracts from the primary treaty aim of climate change mitigation (ibid.: 73). The virtue of a treaty aimed at efficiency is that every state has a reason to endorse it since no state is made worse off and at least one is made better off.

One response to the Posner–Weisbach position might simply seek to augment the goal of an efficient climate regime with the goal of securing the possibility of high human development. The idea would be a familiar one — that there are many points along the curve that represents an (internationally) efficient distribution. We need not be morally indifferent among these points; rather, we should select the one in which developing and least developed countries are securely on the path to high human development. This would combine the virtues of giving each state a reason to endorse the treaty with such dual aims and provide hope for developing and least developed sates.

The problem with this response is that it is not certain that there is a point on the international efficiency curve in which states are securely on the path to high human development. It is possible that finding such a path for every state might require either sufficiently deep emissions reductions by industrialized countries or sufficiently large resource transfers that some of the industrialized countries are not as well off as they might have been in the absence of such reductions and transfers. These are the kinds of requirements that Posner and Weisbach refer to as 'redistributive' (ibid.: 96–97).

The policy goal of safeguarding the possibility of high human development cannot be assumed to be compatible with the goal of international efficiency.

What reasons, then, are there on behalf of the goal of securing the possibility of high human development? One, the attainment of a high level of human development is morally important. The constituent indices of human development, per capita income and health and educational attainments, measure progress in matters towards which people rightfully expect institutions to be attentive. Theories of distributive justice (but typically not measures of human development) are concerned with the distribution of these goods, but they are so concerned because these are goods that people are justified in expecting institutions to promote. Not raising the energy-related costs of human development, then, is a morally important goal of social institutions. It is a goal relevant to climate change policy in particular because an important reason as to why we should care about the effects of climate change is that we care about human development.

The second reason has to do with honouring a prior commitment to respect the right of parties to the UNFCCC to sustainable development. States that have ratified the Convention have committed themselves (in Article 3, paragraph 4) to the right to sustainable development. As I have argued previously, development can be sustainable only if developed countries accept the duty to make very deep emissions reductions or to transfer resources for renewable energy production to developing and least developed countries. Failure of developed countries to respect the right would be unreasonable in light of the prior commitment they made. This is for the general reason that in deliberating about how to coordinate action, especially for morally important purposes, failure to abide by previously agreed upon constraints on deliberative proposals frustrates the deliberative process and shows contempt for the other parties deliberating (Moellendorf 2011).

There are, therefore, two kinds of moral reasons that an international climate change agreement should allow for the possibility of high human development through access to relatively inexpensive sources of energy. The first reason (cited formerly) might be thought of as deep insofar as it involves considerations that are independent of prior agreements. The second reason is shallow in the sense that its force is contingent upon a morally licit prior agreement. The shallowness of the second reason is the source of its limitation but also its strength. The limitation is that the reason to respect the right to sustainable development applies only to signatories of the Convention. Happily enough, that covers all, or nearly all, states; so the limitation makes no real practical difference. The strength of the reason is that it need not appeal to an independent moral value that might be controversial. Parties to a deliberative agreement assume duties to honour that agreement when they commit. There may be a number of morally acceptable excuses available to them for not abiding by the agreement (ibid.). But in the absence of such excuses, failure to abide by the terms of the agreement is wrong regardless of whether we think there are morally important independent reasons to act as the agreement requires. This is similar to the reason that one person has to honour a promise to another. My duty to fulfil my promise to help my neighbour does not depend upon whether there are reasons independent of the promise to help my neighbour.

The argument would be incomplete without considering how these reasons fare in comparison to that for supporting efficiency as the focus of an international climate agreement. A climate change regime that would be reasonably likely to produce an internationally efficient outcome would provide each state with a reason based on prudence or rational self-interest to support it. As such, a proposal for such a regime might be thought to line up with a certain conception of realism about international relations. The idea is that if each state should pursue only an international arrangement that serves its interests, then states would pursue an efficient outcome. But this justification does not, in fact, succeed. An efficient outcome is not necessarily in the best interest of every state. Some powerful states could have bargaining advantages that allowed them to exact concessions out of others that would render them worse off than under a business-as-usual scenario. We need to think only of trade and military relations to imagine how such advantages might be pressed. So, there might be a need to an appeal to a reason, other than prudence, for a state to endorse an agreement that is in everyone's interest.

Stability is not a necessary feature of a consensus generated by mutual self-interested actors, if parties have very unequal power.

To generate such stability in these conditions, other reasons (moral in nature) may be necessary. And, if an efficient regime might have to appeal to reasons other than prudence to motivate stable acceptance, it is unclear why reasons of this other kind would settle necessarily on an efficient regime. An appeal to moral grounds can provide powerful states with grounds to enter into agreements that do not maximally serve their self-interest. In that case, there is no obvious reason why moral reasons should focus on an efficient regime at the expense of all other considerations. And, importantly, the deep and shallow reasons offered earlier for including the goal permitting the possibility of human development provide reasons to reject treaty proposals that threaten human development.

An efficient regime is not necessarily likely to generate a more stable agreement than one that preserves the possibility for human development. Justification of the latter can appeal to both deep and shallow moral reasons to curb national self-interest. But the shallow reasons (discussed previously), for respecting the right to sustainable development, seem to be especially important. The fact that states have already agreed to honour the right to sustainable development is an important moral reason for why they should. And it has the great virtue of being non-controversial. There simply is no reasonable international political morality that holds that agreements should not be honoured, except perhaps in emergency cases.

Conclusion

The ability-to-pay principle of responsibility seems to be superior to principles that would have responsibility tracking historic emissions. Human development is morally important because its constituent elements are so. This is one vital source of moral concern about climate change and climate change policy. Assigning responsibility to safeguard the possibility of human development on the basis of the ability-to-pay principle would either assign responsibility for very deep emissions reductions to the most highly human-developed countries, with developing and least developed countries having access to continued use of relatively cheap fossil fuels; or assign the responsibility for resource, technology or intellectual property transfers to the most highly human-developed countries so that developing and least developed countries might maintain access to relatively cheap sources of energy while reducing their greenhouse gas emissions.

References

Baer, Paul, Tom Athanasiou, Sivan Kartha, and Eric Kemp-Benedict. 2008. *The Greenhouse Development Rights Framework: The Right to Development in a Climate Constrained World*, Heinrich Böll Foundation, Christian Aid, EcoEquity and the Stockholm Environment Institute, Berlin.

Caney, Simon. 2005. 'Cosmopolitan Justice, Responsibility, and Global Climate Change', *Leiden Journal of International Law* 18(4): 747–75.

Carbon Mitigation Initiative. 'Stabilization Wedges Introduction', http://cmi.princeton.edu/wedges/intro.php (accessed 1 December 2011).

Intergovernmental Panel on Climate Change (IPCC). 2007. *Climate Change 2007: Synthesis Report: Summary for Policy Makers*, http//:www.ipcc.ch/pdf/assessment-report/ar4/syr/ar4_syr_spm.pdf (accessed 1 December 2011).

Moellendorf, Darrell. 2011. 'The Right to Sustainable Development', *The Monist* 94(3): 433–52.

———. 2009a. 'Treaty Norms and Climate Change Mitigation', *Ethics and International Affairs*, 23(3): pp. 247–65, http://www.carnegiecouncil.org/publications/journal/23_3/features/001 (accessed 28 April 2013).

———. 2009b. *Global Inequality Matters*, Palgrave Macmillan, Basingstoke.

Neumayer, Eric. 2000. 'In Defence of Historical Accountability for Greenhouse Emissions', *Ecological Economics* 33(2): 185–92.

Posner, Eric A. and David Weisbach. 2010. *Climate Change Justice*, Princeton University Press, Princeton.

Shue, Henry. 1993. 'Subsistence Emissions and Luxury Emissions', *Law & Policy* 15(1): 39–59.

Stern, Nicholas. 2007. *Stern Review: The Economics of Climate Change*, Cambridge University Press, Cambridge, http://webarchive.nationalarchives.gov.uk/+/http://www.hm-treasury.gov.uk/d/Chapter_9_Identifying_the_Costs_of_Mitigation.pdf (accessed 1 December 2011).

United Nations Development Programme (UNDP). 2007. *Human Development Report 2007–2008*, Palgrave Macmillan, Basingstoke, http://hdr.undp.org/en/media/HDR_20072008_EN_Complete.pdf (accessed 20 October 2011).

United Nations (UN). 1988. *General Assembly, Resolution 43/53*, http://www.un.org/ga/search/view_doc.asp?symbol=A/RES/43/53&Lang=E&Area=RESOLUTION (accessed 28 April 2013).

Vanderheiden, Steve. 2008. *Atmospheric Justice: A Political Theory of Climate Change*, Oxford University Press, Oxford.
Weisbach, David. 2010. 'Negligence, Strict Liability, and Responsibility for Climate Change', *The Harvard Project on International Climate Agreements*, Discussion Paper 1–39, http://belfercenter.ksg.harvard.edu/files/WeisbachDP39.pdf (accessed 20 October 2011).

6

Climate Change and Developing Countries: From Leadership to Liability

Joyeeta Gupta

With the 2011 decision of Canada and Russia to join the United States of America (USA) in taking a more passive role in global climate change politics, climate change negotiations run the risk of moving to the back-burner of international politics. Developed countries that had, in the 1992 United Nations Framework Convention on Climate Change (UNFCCC), promised to 'lead' the world by finding a way to both reduce their own emissions of greenhouse gases and help the developing countries to find a short-cut to sustainable development, through the use of technology transfer and financial assistance (Gupta 1998), are after a long period of 'conditional leadership', slowly beginning to renegade on their promises. The European Union (EU) now stands alone with an unconditional offer to reduce its emissions; but this offer may come under pressure as a consequence of the recession and the Euro crises.

Is this trend of failing leadership acceptable? Some academics argue, first, that in an anarchic and rational actor world, it is only logical that countries will tend to free ride on global public goods issues (Posner and Sunstein 2010; Stavins 2011); and that there is no overriding reason for co-operation that requires far-reaching sacrifice from some countries. Second, such sacrifice is inevitable as the Environmental Kuznets curve does not hold for global emissions of greenhouse gases and that decoupling economic growth from emissions is very difficult; a point confirmed by a recent United Nations Development Programme (UNDP) report (UNDP 2011).

However, I would argue that climate change requires global co-operation because (*a*) it is a global problem; (*b*) co-operation helps to bring together the best of global science and knowledge to address the problem and help determine global thresholds; (*c*) co-operation helps to allocate rights and responsibilities between countries in a legitimate manner in order to promote the common good; (*d*) it allows for global offset mechanisms (such as the Clean Development Mechanism or possibly Reducing Emissions from Deforestation and Forest Degradation, in its offset variant) which would otherwise be impossible, as well as other co-operative mechanisms; and (*e*) it provides the only possible venue for countries that are most vulnerable to climate change to make their claims heard.

However, many countries, especially those with high (past, current and, possibly, future) emission levels prefer to argue against multilateral co-operation as they wish to promote their own national interests, avoid liability for damage caused elsewhere, use divide and control or inclusion and exclusionary strategies, and argue that given the 'free-rider' nature of the problem it is more effective to mobilize local populations to voluntarily contribute to problem solving (Gupta 2008).

Problem definition is key to identifying solutions to a problem (Hisschemöller 1993). In the past, the climate change problem has often been defined as a technological one (requiring technological solutions) or as a market problem (requiring economic flexibility mechanisms). However, it is increasingly becoming clear that neither technologies nor markets will solve the problem alone.

In a previous article (Gupta 2012), I argued that the application of four different theories to the climate change problem leads to the conclusion that the latter is essentially a normative problem, that is, a problem where the main issue is the lack of agreement about the values underlying how the problem is to be addressed. If one diagnoses it from a 'problem structuring' approach, it becomes clear that the lack of consensus on climate change is not so much about the science but about the values. If one examines the problem using negotiating theory, it is clear that soft constructive bargaining is needed, one that takes the normative issues into account. If one looks at the problem using a 'social practice' lens, then there is a need to examine what is normatively right in order to address the problem. And finally, a cornerstone of international law is the idea of a common normative base and that norms help to shape behaviour. The question is: do countries first develop customs and then adopt norms, which represent these customs, or do they first adopt norms and then behave in conformity with those norms? Historically, the former was more significant but as we move into a world of rapid change, the latter approach may make more sense. The latter is referred to as 'normative force' in the literature (van Dijk 1987). I concluded the 2012 article by arguing that there is need for leadership on the promotion of norms regarding how resources, rights, responsibilities, and risks are to be shared between countries through the identification of an appropriate social practice framing of problems and by encouraging constructive and integrative bargaining within the multilateral United Nations (UN) legal order.

This chapter goes a step further to address two questions: how legitimate is it to continue to expect and demand leadership from the developed countries in a changing global context and how can litigation and/or dispute resolution in a variety of contexts on climate change help to create precedents that may have global relevance and help bring the climate change negotiation process to the next level, where it is more capable of addressing global problems? In fact, as science and technology, financial and economic institutions, and communication and transport become more and more integrated, global problems will rise rapidly. Expectations about how the future will develop in terms of demographics, economy and environmental impact make it increasingly important to find ways to develop appropriate institutions to address these problems (Gupta and Sanchez 2012a).

Leadership and the Common but Differentiated Responsibilities and Respective Capabilities (CBDRRC) Principle

In the run-up to climate change negotiations, there were discussions regarding how responsibilities (to deal with the problem) were to be allocated between countries. Although early discussions did look at the possibility of holding countries liable for the harm caused to others, political declarations that followed (such as the Noordwijk Declaration on Climate Change 1989 and the Second World Climate Conference Political Declaration 1990) developed the leadership model that was consolidated in the UNFCCC as follows:

> The Parties should protect the climate system for the benefit of present and future generations of humankind, on the basis of equity and in accordance with their common but differentiated responsibilities and respective capabilities. Accordingly the developed countries should *take*

> *the lead* in combating climate change and the adverse effects thereof (1992: Preambular, paragraph 1, emphasis mine).

Leadership is an attractive concept for three reasons. First, it is a positive framing of responsibility as opposed to the use of words such as 'polluter' and/or 'liability'. Second, in an anarchic world, leadership is needed to address collaboration, as opposed to co-operation, problems. Third, if the leadership could lead to the articulation of 'no-regrets' options and technological and market solutions, this would benefit all countries and lead to a win-win situation. Leadership was articulated initially as a combination of emission reduction by the developed countries, the provision of technology transfer and 'new and additional' financial assistance to developing countries (to both reduce the rate of growth of their emissions and help them adapt to the inevitable impacts of climate change), and the provision of room for developing countries to grow. One can argue that the leadership discourse included a number of principles — the precautionary principle, sustainable development, and recognition of the specific needs and special circumstances of the developing countries (UNFCCC 1992, Art. 3.2, 3.3 and 3.4). It also requires, on the one hand, coherence with the open international trading system (UNFCCC 1992, Art. 3.5) and, on the other hand, acceptance of the right to develop of developing countries as embodied in the UN General Assembly Resolution of 1986 (UNGA 1986).

The treatment of the leadership concept has gone through several phases. Initially, in the pre-1992 period, countries in the Organisation for Economic Co-operation and Development (OECD), excluding the USA and Turkey, had adopted national political targets to reduce their emissions. The end of the cold war meant that there would be resources that could now be channelled towards environmental issues and other global challenges. This enthusiasm underlined the agreement framed in the UNFCCC. However, the USA's reluctance to adopt targets led to a weakening of the text on targets (Bodansky 1993), while the text on assistance, although crystal clear in words,

> the extent to which developing country Parties will effectively implement their commitments under the Convention will depend on the effective implementation by developed country Parties of their

> commitments under the Convention related to financial resources and transfer of technology (UNFCCC 1992: Art. 4.7).

was not articulated in terms of targets and timetables.

It was expected that emission reduction targets would have sharpened over time and that assistance to developing countries would have been spelt out. But by the time the Kyoto Protocol was negotiated, the overall target of a –5.2 per cent reduction in 2008–12, with respect to 1990 levels, was weak as it included targets to increase emissions in some cases (such as Australia, Iceland, Norway, and some EU countries) and allowed for compensating emission increases in the developed world through reductions in the developing world, by means of the Clean Development Mechanism (CDM). It was also clear that the United States (US) Senate was unlikely to support the Protocol (Byrd Hagel Resolution 1996). The leadership had morphed into conditional leadership — the US was waiting for developing countries to take action, and the EU for the US and Japan to take action. In the following phase, the EU ratified the Protocol, which came into force in 2005. This led to leadership competition as the US now tried to develop alternative venues for discussing climate change, leading to a number of different agreements. Academic papers began to focus on possible ways to push the climate change regime further but not necessarily within the UNFCCC regime. Since the recession in 2008, leadership on climate change has been receding. The EU made one last effort at a bold unconditional emission reduction target of 20 per cent by 2020, while developing countries have come up with a series of targets. However, the bottom line is that most of the targets (except the EU target) are conditional on action by others; the process is now in limbo! Meanwhile, discussions have moved to adaptation, forests and nationally appropriate mitigation actions in the developing world,but leadership on targets, timetables and financing is rapidly waning.

The Implications of the Changing North-South Context on the CBDRRC

Leadership and the CBDRRC principle were developed at a time when the world was moving from a tri-polar set-up of first, second and third worlds, to a developed–developing country bi-polar set-up. The question is — the world in which this paradigm was developed is changing; does this make this paradigm less relevant today and does that have any implications on the nature of the values for dealing with climate change?

In the run-up to 1990, the OECD consisted of 26 countries and the EU of 12, and there were around 40 developed countries. When the UNFCCC was being drawn up, these 40 countries were included in Annexure I that included these so-called developed countries.[1] Even at that time, some of the developed countries objected to the fact that they were expected to take on responsibilities to assist developing countries as they were themselves facing economic collapse. These countries, with economies in transition were, thus, able to negotiate that a new list of Annexure II countries should be drawn up of the very rich countries in the world.

However, there were no criteria as to what constituted a developed and a developing country, respectively. In the 20 years that have followed, countries have become richer and poorer. South Korea, Mexico and Chile have become members of the OECD. Singapore and Qatar have long been rich countries. But there are also developed countries with lowering GDP, such as Latvia and Lithuania. In recent years, Brazil, China and India are seen as developing rapidly while the recession has hit Greece, Spain, Ireland, and other countries in the EU. At the same time, growth in many developing countries is unstable. The 'Asian Tigers' and Argentina have seen their ups and downs; recently, in 2012, despite its impressive growth record, India was portrayed as a 'junk' economy in the news. The point is that the world is dynamic and change is inevitable.

The lack of criteria for what constitutes a developed country and what kind of leadership would be expected from it has had a negative effect on the negotiating process. This has led some developed

[1] Australia, Austria, Belarus**, Belgium, Bulgaria, Canada, Croatia*, Czechoslovakia (now, The Czech Republic* and Slovakia*), Denmark, EEC (now, European Community), Estonia, Finland, France, Germany, Greece, Hungary, Iceland, Ireland, Italy, Japan, Latvia, Liechtenstein*, Lithuania, Luxembourg, Monaco*, Netherlands, New Zealand, Norway, Poland, Portugal, Romania, Russian Federation, Slovenia*, Spain, Sweden, Switzerland, Turkey**, Ukraine, the United Kingdom (UK), USA. *Countries added to Annexure I by amendment, adopted at COP 3, that came into force on 13 August 1998; **Countries not in Annexure B of the Kyoto Protocol.

countries to undermine the CBDRRC principle see, for example, the decision of the USA, Canada and Russia to take a back seat in the negotiations of future targets as well as the limited resources that are being made available for climate change mitigation and adaptation.

Let us return to the CBDRRC principle and its link to leadership. This principle was adopted in the UNFCCC (1992) and in the Rio Declaration (1992). It implies that some countries have a higher responsibility because of their emissions or because they have a higher ability to pay. This does not mean that other countries do not have any responsibility.

The academic dispute on the CBDRRC rests on four ideas: first, that differentiation based on territory or population does not help to address the problem as it is gross emissions of sovereign states that are critical (Weisslitz 2002; Adams 2003; Bafundo 2006); second, a static interpretation of the principle is unfair, as many developing countries are progressing rapidly (Gupta 2005); third, the mere size of a handful of countries that can single-handedly affect world emissions makes the CBDRRC approach less relevant; and fourth, making exceptions for poorer countries to allow them to grow creates new problems as they often lack the governance framework to subsequently control their emissions (Bafundo 2006). The concept of differentiation is, thus, being challenged.

At the same time, the concept of differentiation is acquiring some degree of legitimacy. Within the global trade regime, where non-discrimination between trading countries and equal treatment of imported and local goods are key principles. An analysis of six cases before the dispute resolution panel of the World Trade Organization (WTO) shows that under some circumstances differentiation is seen as valid. These insights can be summed up as: first, unilateral trade restrictions to protect the environment may be less preferred to joint approaches with developing countries to find common but differentiated solutions to addressing environmental problems. Second, development, financial and trade needs of developing countries need to be taken into account. Third, differentiation may be justified to help developing countries secure a share in the global market. Finally, however, as developing countries progress, they must gradually relax import restrictions; in other words, they cannot indefinitely claim benefits under the CBDRRC principle (Gupta and Sanchez 2012b).

One can, thus, argue that the CBDRRC principle will only be effective and fair when there is clear criteria underlying the principle; when country classifications are seen as dynamic and countries are allowed to graduate in and out of a specific classification (Matsui 2002); when the CBDR part of the principle is not interpreted as exempting developing countries from all responsibilities (instead, they have a responsibility to try and find alternative growth paths and production and consumption patterns (Matsui 2002; Mumma and Hodas 2008) in order to avoid the structural problem of 'lock-in'); that there is a responsibility of those with capacity to help those without, with 'new and additional' resources that are over and above official development assistance (as this is also seen as a pacification strategy to avoid paying compensation for the harms caused to other countries); and that such resource demands need to be kept within justifiable limits (and not for all environmental problems). Finally, the CBDRRC principle puts all countries on notice — that is, if they graduate into a specific category they will have to take on certain responsibilities (Rajamani 2002; Gupta 2003).

Developing the CBDRRC Concept Further

One can argue that the leadership regime was created to avoid liability. It allowed developed countries to be presented in a positive light (and not as polluters) and appeared to provide a constructive and integrative solution. A leader–follower paradigm is after all a win-win solution. This leadership paradigm was expressed in terms of the CBDRRC principle.

However, one can argue that differentiated responsibilities (CBDR) refers to the notion of the 'no harm' principle — that countries can take action within their territorial boundaries but that such action should not cause harm to others. Respective capabilities (RC) refers to differentiation with respect to the wealth of countries and can be linked to the 'ability-to-pay' principle — which, for example, underlies differential taxation within countries.

In terms of the no-harm principle, one can argue that there is action that can be taken 'before' and 'after'. The 'before' action includes the adoption of the precautionary principle and of strict environmental standards, the implementation of environmental impact assessments, the notification of planned measures that may lead to higher emissions, and the adoption of the polluter or user pays principle. All these measures are expected to have an impact on reducing the growth rate of emissions. In terms of the 'after', once environmental harm has occurred, the options include assessing liability, various types of injunctive relief to stop the polluting activity, payment of compensation to actors affected negatively by the action, and the principle of 'allocation of loss'.

In terms of the ability-to-pay principle, one can differentiate between the rich and the poor. For the rich, this implies possibly using higher technical standards and helping poorer countries with capacity building, technology transfer and financial assistance. For the poorer countries, there are lower regulatory standards or differentiated care responsibilities, and these countries are allowed to prioritize poverty eradication and the right to development. In addition, there is the possibility to insure against possible problems.

Litigation as a Way to Articulate Values

Political processes are short-term processes; governments come to power for only a period of four to five years. This is not synchronized with the nature of environmental problems which have a much longer time-frame. Therefore, the institutional fit between the cabinets of countries and climate change is weak and governments will be tempted to choose for short-term, cost-effective options over long-term responsibility.

This is where the role of courts and the judiciary becomes increasingly more important. As norms are gradually being developed at the global level, there needs to be some kind of enforcement mechanism to ensure that these norms are respected. Litigation or preparation for litigation could be one way forward.

Although developing countries have for long demanded that developed countries assume their responsibilities (Gupta 1998), academic interest in climate litigation as a potential subject grew only in the last decade, supported by a growing social movement on climate justice (Peñalver 1988; Marburg 2001; Allen 2003; Grossman 2003; Verheyen 2003; Burns 2004; Gillespie 2004; Thackeray 2004; Hancock 2005; Jacobs 2005; Mank 2005; Gupta 2007b).

There have already been a number of court cases around the world on issues related to climate change. These cover the issue of whether export credits provided by developed countries for investments in the developing world have led to increased greenhouse gas emissions or not. These cases have argued in favour of gathering information about such projects through implementing the freedom of information acts operating in the country (for example, Germany) or implementing environmental impact assessments on such investments (for example, the USA). They have argued that mining expansions and other such investments were undertaken in breach of environmental impact assessments (for example, in Australia). There have been arguments that gas flaring is a violation of human rights and exacerbates greenhouse gas emissions (for example, in Nigeria). Other cases have argued that power companies and their emissions cause a common nuisance and that CO_2 emissions should be seen as a pollutant that falls under environmental impact assessments (for example, in the USA). Meanwhile, some countries have requested the United Nations Educational, Scientific and Cultural Organization (UNESCO) to list sites within their territories as World Heritage Sites in order to offer these some protection (for example, Belize and Peru). Some countries have pushed the UN General Assembly (UNGA) to adopt the Human Right to Water and Sanitation in 2010 (and this is arguably because they feel that their water supplies will be affected by climate change).

At the same time, the legal literature explores a number of alternative options, seeking advisory opinions from the International Court of Justice about the 'good faith nature' of climate change convention negotiations (Gillespie 2004); whether the precautionary and equity principles are being implemented or not (Jacobs 2005); whether the principle of state responsibility and international liability is supportive of far-reaching action ór not (Verheyen 2003); and whether the non-ratification of the Kyoto Protocol by the USA is *de facto* an illegal subsidy to companies under the WTO (Burns 2004; Doelle 2004), amongst many others.

One can argue that there is high agreement in the legal literature that failure or stagnation of the international negotiating process can be countered by demanding the accountability of governments, industry and individuals through the use of legal remedies (Peñalver 1998; Allen 2003; Grossman 2003; Verheyen 2003; Burns 2004; Doelle 2004; Gillespie 2004; Hancock 2005; Jacobs 2005; Smith and Shearman 2006). The interest in and use of legal action is continuously increasing. Haritz (2010) has argued that based on an analysis of the literature and court cases, it is theoretically possible to link the Intergovernmental Panel on Climate Change (IPCC) scale of likelihood with a scale based on legal standards of proof required for different kinds of legal action. For example, IPCC's 'virtually certain' can correspond to 'beyond a reasonable doubt' and could apply for a criminal conviction case; while 'very likely' and 'likely' correspond to 'clear and convincing evidence' for a quasi-legal penal action and a 'clear showing' that could be relevant in a temporary injunction, respectively. Liability for climate change damage at the supranational level (Gouritin 2011; de Larragán 2011; Peeters 2011) and at the national level in the UK (Kaminskaite-Salters 2011), the USA (Kosolapova 2011), and the Netherlands (van Dijk 2011) is increasingly being explored as an option. Legal literature, thus, sees climate litigation and legal liability as possible, as being already used in courts, and as likely to put additional pressure on corporations and governments to be more accountable in the future (Smith and Shearman 2006; Faure and Peeters 2011). There are, however, familiar hurdles (such as, can the cause–effect be proved, who can be held responsible for the cause, which court is relevant for a specific case, and so on) and the law will have to develop further in this field.

Conclusion

This chapter has tried to look at leadership and the principle of common but differentiated responsibilities and respective capabilities. It has raised a number of questions, which I will now try and answer in an integrated manner. Is the CBDRRC principle still legitimate in a changing world? Yes it is: as long as there continues to be major differences between the rich and the poor, the need to differentiate between countries is principally fair (UNDP 2011). Should developed countries take action first? Yes, since most of the current and near future impacts are caused by emissions by developed countries and by countries with economies in transition in the past. The growing emissions of developing countries will exacerbate climate change in the future and a good precedent needs to be set in order to expect them to change their behaviour accordingly.

Having said that, how does one counter the arguments that the CBDRRC is *de facto* neither fair nor environmentally effective?

The answer to the first question is that the CBDRRC must be elaborated to include criteria for classifying countries, allow for dynamic movement of countries from one category to another, give differentiated tasks to each group of countries, and ensure that no group is exempted from the responsibility of trying to find a path towards sustainable development. Furthermore, given that a handful of countries have higher gross emissions, co-operation between these countries needs to be further intensified in search of affordable emission reduction options as well as developing pathways that avoid institutional and technological lock-in. However, this needs to be undertaken through soft constructive bargaining processes leading to win-win options, rather than hard distributive bargaining, which tends to have poor results.

A third question raised was: is this just a normative idea or can it be operationalized further? This chapter argues that it can be operationalized further into its components and can be developed building on existing policy and legal principles and tools. Such operationalization is necessary to make the concept stronger.

The fourth question was: given the reluctance of countries to agree on norms at the global level, how can accountability be developed further? This chapter claims that short-term democracy implies that cabinets across the world may be unwilling to take action that requires sacrifices. However, the institutions of democracy that are more 'stable' — the judiciary, for instance — could rise above short-term politics. Using litigation (at a local as well as a global level) is a viable option, which is gathering momentum in the literature and in courts.

References

Adams, Todd B. 2003. 'Is there a Legal Future for Sustainable Development in Global Warming? Justice, Economics and Protecting the Environment', *Georgetown International Environmental Law Review* 16(1): 77–126.

Allen, Myles. 2003. 'Liability for Climate Change: Will it Ever Be Possible to Sue Anyone for Damaging the Climate?', *Nature* 421: 891–92.

Bafundo, Nina E. 2006. 'Compliance with the Ozone Treaty: Weak States and the Principle of Common but Differentiated Responsibilities', *American University International Law Review* 21(3): 461–95.

Bodansky, Daniel. 1993. 'The United Nations Framework Convention on Climate Change: A Commentary', *Yale Journal of International Law* 18(2): 451–588.

Burns, William G.C. 2004. 'The Exigencies that Drive Potential Causes of Action for Climate Change Damages at the International Level', *American Society of International Law Proceedings* 98: 223–27.

Byrd Hagel Resolution 1996. Byrd Hagel Congressional Record: 3 October 1997 (US senate), S10308–10311.

de Larragán, Francisco Javier de Cendra. 2011. 'Liability of Member States and the EU in View of the International Climate Change Framework: Between Solidarity and Responsibility', in Michael Faure and Marjan Peeters (eds), *Climate Change Liability*, Edward Elgar, Cheltenham Gloss, 55–89.

Doelle, Meinhard. 2004. 'Climate Change and the WTO: Opportunities to Motivate State Action on Climate Change through the World Trade Organization', *Review of European Community and International Environmental Law* 13(1): 85–103.

Gillespie, Alexandre. 2004. 'Small Island States in the face of Climate Change: The End of the Line in International Environmental Responsibility', *UCLA Journal of Environmental Law and Policy* 22(1): 107–29.

Gouritin, Armelle. 2011. 'Potential Liability of European States under the ECHR for Failure to Take Appropriate Measures with a View to Adaptation to Climate Change', in Michael Faure and Marjan Peeters (eds), *Climate Change Liability*, Edward Elgar, Cheltenham Gloss, 134–64.

Grossman, David A. 2003. 'Warming up to a Not-so-Radical Idea: Tort Based Climate Change Litigation', *Colombia Journal of Environmental Law* 28(1): 1–62.

Gupta, Joyeeta. 2012. 'Negotiating Challenges and Climate Change', *Climate Policy*, 12(5): 630–44.

———. 2008. 'Global Change: Analysing Scale and Scaling in Environmental Governance', in Oran R. Young, Heike Schroeder and Leslie A. King (eds), *Institutions and Environmental Change: Principal Findings, Applications, and Research Frontiers*, MIT Press, Massachusetts, 225–58.

———. 2007. 'Legal Steps Outside the Climate Convention: Litigation as a Tool to Address Climate Change', *Review of European Community and International Environmental Law* 16(1): 76–86.

———. 2005. 'International Law and Climate Change: The Challenges Facing Developing Countries', *Yearbook of International Environmental Law* 16(1): 119–53.

———. 2003. 'Engaging Developing Countries in Climate Change: (KISS and Make-Up!)', in David Michel (ed.), *Climate Policy for the*

21st Century: Meeting the Long-Term Challenge of Global Warming, Johns Hopkins University Press, Washington, 233–64.

Gupta, Joyeeta. 1998. 'Leadership in the Climate Regime: Inspiring the commitment of developing countries in the post-Kyoto phase', *Review of European Community and International Environmental Law* 7(2): 178–88.

Gupta, Joyeeta and Nadia Sanchez. 2012a. 'Global Green Governance: The Green Economy Needs to be Embedded in a Global Green and Equitable Rule of Law Polity', *Review of European Community and International Environmental Law*, 21(1): 12–22.

———. 2012b. 'Elaborating the Common but Differentiated Responsibilities Principle: Experiences in the WTO', Legal Working Paper, International Developmental Law Organization, Rome, http://www.idlo.int/Documents/Rio/05.%20CBDR%20Experiences%20in%20the%20WTO.pdf (accessed 31 March 2014).

Hancock, Elizabeth E. 2005. 'Red Dawn, Blue Thunder, Purple Rain: Corporate Risk of Liability for Global Climate Change and the SEC Disclosure Dilemma', *Georgetown International Environmental Law Review* 17(2): 223–51.

Haritz, Miriam. 2010. *An Inconvenient Deliberation: The Precautionary Principle's Contribution to the Uncertainties Surrounding Climate Change Liability*, PhD Thesis, Maastricht University, Box Press Publishers, Oisterwijk.

Hisschemöller, Matthijs. 1993. *De democratie van problemen. De relatie tussen de inhoud van beleidsproblemen en methoden van politieke besluitvorming*, Vrije Universiteit, Amsterdam.

Jacobs, Rebecca E. 2005. 'Treading Deep Waters: Substantive Law Issues in Tuvalu's Threat to Sue the United States in the International Court of Justice', *Pacific Rim Law and Policy Journal* 14(1): 103–28.

Kaminskaite-Salters, Giedre. 2011. 'Climate Change Litigation in the UK: Its Feasibility and Prospects', in Michael Faure and Marjan Peeters (eds), *Climate Change Liability*, Edward Elgar, Cheltenham Gloss, 165–88.

Kosolapova, Elena. 2011. 'Liability for Climate Change-related Damage in Domestic Courts: Claims for Compensation in the USA', in Michael Faure and Marjan Peeters (eds), *Climate Change Liability*, Edward Elgar, Cheltenham Gloss, 189–205.

Mank, Bradford C. 2005. 'Standing and Global Warming: Is Injury to All, Injury to None?', *Environmental Law* 35(1): 1–83.

Marburg, Kristin L. 2001. 'Combating the Impacts of Global Warming: A Novel Legal Strategy', *Colorado Journal of International Environmental Law and Policy*, 2001 Yearbook: 171–80.

Matsui, Yoshiro. 2002. 'Some Aspects of the Principle of "Common but Differentiated Responsibilities"', *International Environmental Agreements: Politics, Law and Economics* 2(2): 151–70.

Mumma, Albert and David Hodas. 2008. 'Designing a Global Post-Kyoto Climate Change Protocol that Advances Human Development', *Georgetown International Enviromental Law Review* 20(4): 619–43.

Noordwijk Declaration on Climate Change. 1989. *Noordwijk Conference Report*, Volume I, ed. by Piers Vellinga et al., The Hague.

Peeters, Marjan. 2011. 'The Regulatory Approach of the EU in View of Liability for Climate Change Damage', in Michael Faure and Marjan Peeters (eds), *Climate Change Liability*, Edward Elgar, Cheltenham Gloss, 90–133.

Peñalver, Eduardo M. 1998. 'Acts of God or Toxic Torts? — Applying Tort Principles to the Problem of Climate Change', *Natural Resources Journal* 38(4): 563–69.

Posner, Eric and Cass Sunstein. 2010. 'Justice and Climate Change: The Unpersuasive Case for Per Capita Allocations of Emissions Rights', in Joseph E. Aldy and Robert N. Stavins (eds), *Post-Kyoto International Climate Policy*, Cambridge University Press, Cambridge, 343–71.

Rajamani, Lavanya. 2002. *Differential Treatment in International Environmental Law: Sharing the Burden of Climate Protection*, Phd Thesis, Faculty of Law, University of Oxford, Oxford.

Rio Declaration 1992, http://www.unep.org/Documents.Multilingual/Default.asp?documentid=78&articleid=1163 (accessed 16 April 2014).

Second World Climate Conference. 1990. *Ministerial Declaration of the Second World Climate Conference*, Geneva.

Smith, Joseph and David Shearman. 2006. *Climate Change Litigation: Analysing the Law, Scientific Evidence and Impact on the Environment, Health and Property*, Presidian Legal Publications, Adelaide.

Stavins, Robert N. 2011. 'The Problem of the Commons: Still Unsettled After 100 Years', *American Economic Review* 101(1): 81–108.

Thackeray, Richard W. 2004. 'Struggling for Air: The Kyoto Protocol, Citizen's Suits Under the Clean Air Act, and the United States Options for Addressing Global Climate Change', *Indiana International and Comparative Law Review* 14(3): 855–903.

United Nations Development Programme (UNDP). 2011. *Human Development Report, Sustainability and Equity: A Better Future for All*, Palgrave Macmillan, Basingstoke.

United Nations Framework Convention on Climate Change (UNFCCC). 1992. *United Nations Framework Convention on Climate Change*, New York, 9 May 1992; in force on 24 March 1994.

UN General Assembly (UNGA). 1986. *Declaration on the Right to Development*, United Nations General Assembly Resolution A/RES/41/128, 4 December 1986.

Van Dijk, Chris. 2011. 'Civil Liability for Global Warming in the Netherlands', in Michael Faure and Marjan Peeters (eds), *Climate Change Liability*, Edward Elgar, Cheltenham Gloss, 206–26.

Van Dijk, Pieter. 1987. 'Normative Force and Effectiveness of International Norms', *German Yearbook of International Law* 30: 9–35.
Verheyen, R. 2003. 'Climate Change Damage in International Law', Dissertation zur Erlangung des Dr. iur Universitat Hamburg, Fachbereich Rechtswissenschaft, Hamburg.
Weisslitz, Michael. 2002. 'Rethinking the Equitable Principle of Common but Differentiated Responsibility: Differential versus absolute norms of compliance and Contribution in the Global Climate Change Context', *Colorado Journal of International Environmental Law and Policy* 13: 473–509.

Part III

Topics in Moral Theory

7

Contractualism and Climate Change

Jussi Suikkanen

Climate change is 'a complex problem raising issues across and between a large number of disciplines, including physical and life sciences, political science, economics, and psychology, to name just a few' (Gardiner 2006: 397). It is also a moral problem. Therefore, in this chapter, I will consider what kind of contribution an ethical theory called 'contractualism' can make to climate change debates.

I will first introduce contractualism. I will then describe a simple climate change scenario. The third section explains what kind of moral obligations we would have in that situation according to contractualism. Finally, in the last section, I will discuss some of the advantages and problems of the sketched view. These discussions should help us better understand contractualism and illustrate how it could, perhaps, enable us to come to grips with some of the more difficult moral aspects of climate change.

Contractualism

I will focus here on contractualism as explained in T. M. Scanlon's *What We Owe to Each Other* (1998). It can be understood as an answer to the questions 'Which actions are wrong?' and 'Why should we not do those actions?' According to contractualism, an act is wrong whenever it is forbidden by the set of moral principles which no one could reasonably reject (ibid.: 4). Therefore, we first need define the set of principles that could not be reasonably rejected.

Imagine a set of possible worlds that are like our own world. The only difference between these worlds and ours is that different sets of moral principles have been internalized in them. For any set of moral principles we could adopt, there is a possible world in which it has already been adopted.

The way in which agents behave in these worlds is influenced by the internalized moral principles. Therefore, in the worlds where the principle 'do not kill' has been internalized, killings rarely happen. Likewise, in the world where drinking coffee violates a moral principle, no coffee is drunk. Of course, in most worlds, the adopted rules are much more sophisticated than these hypothetical illustrations. In those worlds, the previous actions are done in some contexts but not in others.

The actions that are done in these worlds determine what kind of lives people come to live in them. They create personal 'standpoints' of the individuals living in these worlds (Scanlon 1998: 202–6). In some worlds, many individuals live happier lives because help is offered to strangers there. In other worlds, people fail to have close personal relationships because the moral codes of those worlds require treating everyone equally, in a very literal way. Some features of personal standpoints in these worlds count as 'burdens'. For instance, not being able to form close personal relationships is a burden because it is a feature that makes one's life less choice-worthy.

We can then define when a set of moral principles cannot be reasonably rejected (ibid.: 195). Any individual in the previous worlds can reject the set of principles under which she lives, when she can point to another world in which no one needs to experience serious burdens like hers. Thus we first find, from each world, the individual who has to live the most burdensome life in that world. We then ask which individual (of those individuals) lives the least burdensome life. The set of moral principles under which this person lives is the one which no one could reasonably reject. This is because all other sets of moral principles cause more serious personal burdens to some individuals. The individuals who have to bear those burdens can reasonably reject the principles under which they live because of the way in which they are unnecessarily burdened. By 'unnecessary burdens' I merely mean that no one needs to experience qualitatively similar burdens in the other worlds that are being compared.[1]

---

[1] I also assume here that individuals are 'worldbound' and so no one lives in more than one of these compared worlds. Hence, we are always comparing pair-wise the concrete standpoints of individuals a, b, . . ., n to the standpoints of different individuals r, s, . . ., z. Furthermore, these

The second part of contractualism explains why we should follow the non-rejectable principles (ibid.: ch. 4, esp. p. 162). By following them, we can form relationships with other people such that in them we mutually recognize each other's ability to evaluate and act on reasons. These relationships are valuable for the following reasons.

If we fail to act in ways that can be justified to others on non-rejectable grounds, we express our willingness to ignore the objections they might have to our actions. This would belittle their abilities to evaluate reasons. We too would be offended if others did not care about whether they could justify their actions to us. In contrast, if we are able to justify our actions on the non-rejectable grounds, we can stand by our actions and proudly look into the gaze of others without feeling like shrinking (Pettit 2000: 231). This relation to others constitutes a concrete good in our lives.

After this brief summary of contractualism, the question is whether this framework could be used to determine how we should react to climate change.

A Climate Change Scenario

Public debates about climate change are often about empirical facts. People disagree about (*a*) whether the continued emission of the greenhouse gases will change Earth's climate, and about (*b*) how the projected change in climate will affect life on Earth. As a moral philosopher, I am unqualified to contribute to these debates. For more information, it would be much better to consult the latest scientific report by the Intergovernmental Panel on Climate Change (IPCC 2013).

However, when we study moral philosophy, we can often ignore the empirical questions because when we focus on purely ethical questions, we can stipulate the facts of relevant situations and then ask what we should do in them. Thus, when we study ethics, we

compared individuals cannot present their non-existence in the other worlds as objections to those other sets of principles but rather only the undesirable concrete qualities of their actual lives to the ones under which they live (see Scanlon 1998: 204). I've argued elsewhere that this framework does not lead to the repugnant conclusion. It can also explain why principles that lead to very small populations can still be reasonably rejected (Suikkanen 2011).

can go to other possible worlds of which we know all the basic non-moral facts by stipulation. We can then consider what obligations we would have in them. In this way, the empirical facts of the actual world become irrelevant for our ethical investigation.

Here, I will focus on a possible world which contains a planet that, in many ways, at least seems to resemble our own Earth. Call it 'Earth*'. On Earth*, the global average surface temperature was 13.8°Celsius (C) in the year 1900. The average temperature on Earth* then steadily rose so that it was already 14.6°C in the year 2010.

Because we are stipulating the facts of this world, we know that, on Earth*, this rise in temperature was caused by an increased use of fossil fuels. Burning them emitted CO_2, which functioned as a greenhouse gas by reflecting outgoing radiation back on the planet's surface. So, in this world, climate change is anthropogenic. Thus, on Earth*, climate change can be explained by the fact that the pre-industrial concentration of CO_2 in atmosphere was 280 parts per million whereas it was about 430 parts per million in 2010.

The higher temperatures were already starting to affect the lives of the people who lived on Earth* in the year 2010. Because of the changed climate, these individuals were experiencing more severe weather events: floods, droughts, hurricanes, tornadoes, and so on. More individuals were also dying of heatstroke during severe heat waves. What should the people of Earth* do at this point?

Contractualists answer this question by considering two possible futures of Earth*, which are different because people have adopted different sets of moral principles in them. This leads them to act in different ways. These adopted principles also include climate change policies. They govern all the actions that together determine how much CO_2 and other greenhouse gases will be emitted each year. These actions include both economic and environmental policies set by the governments and the ordinary everyday actions of ordinary people worldwide.

In one future of Earth*, Future A, everyone adopts moral principles that do not change people's behaviour. Call this the adoption of the 'Do Nothing Policy'. As a consequence of this, the atmospheric concentration of CO_2 increases by three parts per million each year. Hence, in Future A, the concentration of CO_2 will be 700 parts per million in the year 2100. This raises the global average temperature to 19°C by the same year. In Future B of Earth*, everyone adopts moral principles that lead to (*a*) drastic cuts in energy consumption, and (*b*) investment in and adoption of carbon-free energy sources. Call this the adoption of the 'Mitigation Policy'. Because that policy is adopted in this future, the emissions of the greenhouse gases are reduced so sharply that the atmospheric concentration of CO_2 is stabilized and eventually even reduced slightly. This, of course, first stabilizes the temperature of Earth* and then begins to slowly lower it.

Achieving this goal requires reducing the CO_2 emissions more than 60 per cent below the 1990 emissions levels (Shue 1993: 41). In Future B, these drastic cuts are done during a fifteen-year period between 2010 and 2025. This means that, at some point after the year 2025, the atmospheric concentration of CO_2 peaks at 470 parts per million. As a result, Earth*'s average temperature rises only by two degrees, to 16.6°C in the year 2040, and then begins to slowly decrease back towards the original temperature.

I will first focus here only on the choice between these two possible climate policies. Between them, there is of course a whole spectrum of more moderate climate policies. However, once we understand how the contractualist framework makes a choice between the two extreme alternatives, we will be able to use the same pairwise assessment-method to evaluate the other policies as well.

A Contractualist Proposal

Contractualists will then want to know if either the 'Do Nothing Policy' or the 'Mitigation Policy' could not be reasonably rejected. This leads them to investigate what kind of personal standpoints these policies create for the individuals who have adopted them in the alternative futures of Earth*.

Let us first consider what kind of lives people will live in Future A under the 'Do Nothing Policy'. Here, the average temperature of the Earth* is 19°C in the year 2100. This does not mean that the temperature on Earth* is uniformly five degrees higher. Some areas will be affected more than others.

The main change in climate is that there are more severe weather events in this future.[2] Because of the more frequent heat waves,

[2] For a description of similar consequences of climate change on the actual Earth, see Caney (2009: sec. 2).

more people will die of heatstroke. Many other severe weather events are water-related. Some areas will experience severe droughts whereas others will flood because of the increased levels of rain. Both disasters will claim human victims directly. They will also spread fatal and painful diseases, such as malaria, diarrhoeal ailments and dengue.

Higher temperatures will mean that the snow cover and ice extent will decrease, making sea levels rise. As a result, many coastal populations will lose their homes and livelihoods. Higher temperatures and the severe water-related weather events will also be bad news for farmers. Lower crop yields will cause hunger and starvation, causing more people to emigrate. All of this will lead to conflicts and wars over scarcer resources.

Therefore, we know that billions of lives in Future A will be terrible. Many will die, suffer from painful diseases, starve, lose their homes and livelihoods, be forced to migrate, and to fight wars.

Things change in the alternative Future B for the worse too. First, the temporary two-degree increase in the Earth*'s average temperature will cause some severe weather events. Some additional people die of heatstroke. There will also be more floods and droughts, and many individuals will need to leave their homes because of higher sea levels. These will be unavoidable because the 'Mitigation Policy' was not adopted before 2010.

Second, individuals who live in this future will also have to bear other types of perhaps less serious personal burdens. In Future B, the radical steps taken to cut greenhouse gas emissions will affect people's everyday lives: cars and airplanes will be smaller and can only be used for most essential travel; houses have to be insulated; use of air-conditioning will be restricted even in hot areas; availability of cheap consumer goods will be constrained; having many children is no longer permitted; food will be sourced locally; meat will not be generally available; fewer careers will be open in energy-heavy industries; and so on.

All of these make most people's lives more burdensome. However, as personal burdens, these life-changes are not as serious as the burdens which people need to bear under the 'Do Nothing Policy' in Future A. Being able to fly less just does not compare to losing one's home because of a hurricane. In Future B, far fewer lives will be lost and there will be fewer individuals who will need to suffer as much as the people who are worst affected in Future A. The only additional burdens which people need to bear in this future are caused by the fact they can no longer live their everyday lives in the same polluting way as before.

Contractualism then asks us to consider whether individuals in either future could reasonably reject the principles under which they live. They can do so if they can point to alternative principles under which no one needs to bear equally serious burdens.

It seems that some individuals in Future A can reasonably reject the 'Do Nothing Policy' whereas the individuals of 'Future B' cannot reasonably reject the 'Mitigation Policy'. Take all the individuals who bear the most serious personal burdens in Future A (premature death, suffering from tropical diseases, starvation, forced migration, and so on). We can pair some of these individuals with each individual who has to experience equally serious burdens in Future B. However, once we have done this, we are left with a group of individuals from Future A, so that (*a*) the lives of these individuals are extremely burdensome, and (*b*) no person from Future B (who has not already been paired with someone from Future A) has to bear equally serious burdens. All the remaining individuals of Future B only need to cope with the changes to their everyday lives caused by the adoption of more environmentally-sound lifestyles.

These remaining burdened individuals of Future A, thus, can reasonably reject the 'Do Nothing Policy' under which they live.[3] That policy creates extremely serious personal burdens to some individuals such that no one needs to bear them under the alternative 'Mitigation Policy'. In other words, if the 'Do Nothing Policy' were followed on Earth*, serious unnecessary burdens would be caused to some individuals. As a result, the actions authorized by that policy could not be justified to everyone else on grounds they could not reasonably reject.

The individuals who happen to suffer under the 'Mitigation Policy' in Future B could not, in the same way, reasonably reject the policy under which they live. We can take every individual

[3] For a more general discussion, see Scanlon (1998: 232–33).

who has to bear extremely serious burdens in Future B and find a counterpart from Future A. Once we have formed these pairs, there is no one left in Future B who has to bear serious personal burdens such that they remain unmatched in Future A. Everyone else in Future B needs to only bear the less serious burdens caused by the requirement to emit less greenhouse gases. Since people do not have to bear avoidable, unnecessary burdens in Future B, no one could reasonably reject the 'Mitigation Policy'.

There are, of course, many other policies which could be adopted on Earth*, and yet the 'Mitigation Policy' seems to be a good candidate for a non-rejectable climate policy. Other policies that cut emissions less than the 'Mitigation Policy' fail to stabilize the atmospheric concentration of greenhouse gases. Therefore, the global average temperature would keep rising under them, which would only delay the bad consequences of climate change. Therefore, these principles could still be reasonably rejected for the very same reasons as the 'Do Nothing Policy'. This shows how contractualism offers a temporarily neutral solution to the question of intergenerational justice.

There are also climate policies that would cut emissions even more radically than the 'Mitigation Policy'. However, they have two downsides. First, they too fail to prevent all bad consequences of climate change. This is because the emissions produced before the year 2010 would continue to change the atmosphere of Earth* in any case. Second, these policies would probably be very intrusive because they would take away too many opportunities for making personal choices. Given how important personal autonomy is, these policies could thus be reasonably rejected because of their intrusiveness (see Scanlon 1998: sec. 6.2).

To conclude, contractualism picks out the policy that no one could reasonably reject, by comparing pair-wise the burdensome personal consequences of alternative climate policies. This policy then sets the moral standards by which actions, in relevantly similar circumstances, are to be evaluated. If our own planet is like Earth* in all relevant respects, then we will act wrongly unless we take the drastic measures recommended by the 'Mitigation Policy'. If we do not, we will not be able to justify our failure to change our behaviour to all future generations on grounds which they could not reasonably reject.

Additional Issues

In this final section, I will discuss some of the advantages and problems of my proposal.

(*a*) The Non-Identity Problem: The first thing to note is that my proposal avoids the 'Non-Identity Problem' (Parfit 1984: ch. 16). The latter is considered to be a decisive objection to many theories of duties towards future generations. The Non-Identity Problem begins from the plausible thesis that if some policies affect how we behave, then they also affect when new individuals will be conceived. Yet, according to most theories of personal identity, who one is essentially depends on the time of one's conception. If your parents had conceived a child at a different time than when they conceived you, then you would not have existed as yourself but rather as some other person.

This has led many to believe that future generations cannot be harmed by the adoption of climate policies (ibid.: 361–64). It is appealing to think that one harms someone when one makes her worse off than she would have been otherwise.[4] Yet, the adoption of an identity-affecting policy cannot make the life of a future person any worse than what it would have been otherwise. This is because without the adoption of the policy that very individual would not have even existed.

This version of contractualism does not rely on the notion of harm. Here, principles can be reasonably rejected because of the personal burdens which they cause to individuals. These burdens need not necessarily be harms. That is, the relevant objections to the principle need not be based on being worse off than one would have been in some alternative scenario. Rather, these burdens are considered to be concrete features of everyday lives (heat strokes, malaria and so on) that make one's life less choice-worthy (Scanlon 1998: 204). These burdens are only compared to the burdens which some other individuals have to experience under the alternative principles. This is why it is not a problem for contractualism that different individuals exist under different principles.

(*b*) The Relationship-based Reasons: So far, I have focused on contractualism as an account of which climate-changing acts might

[4] See Bayles (1976: 293), Parfit (1984: 374) as well as Hanser (1990: 58).

be wrong. Contractualism is also supposed to be a theory of why we should not do those acts. At this point, the view faces a serious problem. Recall that the contractualist reasons for not acting wrongly are based on the value of the reciprocal relations of mutual recognition. These reasons are based on how important it is for us to be able to stand by our actions when we interact with other agents.

However, we cannot have such kind of relations to future generations. When the time comes for these generations to evaluate our actions, we will have ceased to exist. At that point, it is too late for them to exclude us from any relationships. And, it is not clear why these future assessments should matter to us, especially when we do not even know who the assessors will be.

The contractualist reasons to follow the 'Mitigation Policy' must, therefore, be based on the relationships we can have with the people around us now. Violating the policy reveals that one's moral deliberation has not been shaped by the non-rejectable principles. This shows that one is willing to overlook the objections from future standpoints. This fact about one's deliberation should be a reason for concern. In overlooking the objections of future people, one implicitly judges that those individuals do not deserve to be taken into consideration on the basis of their cognitive and affective capacities.

However, we share all those same capacities. The only difference between the future people and us is that we exist right now. So, if someone ignores the objections of future people to our behaviour, this also tells us that she thinks that we do not deserve to be taken into consideration on the basis of our essential qualities. We are supposed to be taken into account on the basis of our 'presentness'. Yet, from our present perspective, this seems like a demeaning judgement of our qualities as persons. This is why we should be concerned about our relation to people who are not willing to justify their actions to future generations on reasonable grounds.

Thus, if someone violates the 'Mitigation Policy' now, this would be a reason for the other currently existing people to distance themselves from that individual. If moral relationships described by Scanlon are valuable, then that individual would have a reason to avert this distancing move by following the 'Mitigation Policy'. In this way, the contractualist account of our reasons not to act wrongly could, perhaps, be extended to apply to the 'Mitigation Policy'.

(*c*) The Anti-Utilitarian Protection: An interesting feature of contractualism is that the rejectability of climate policies is a function of the pair-wise comparisons of the potential personal burdens. Consequently, unlike utilitarianism, contractualism can protect individuals and small groups (Scanlon 1998: 235).

Consider a climate change scenario in which climate change would only affect the northernmost parts of North America. The only affected groups would be 32 million Canadians and a small Inuit population. The potential climate change in question would make the life of the Canadians somewhat easier. Warmer summers and milder winters would make everyone's lives more comfortable. Crops would grow quicker, less energy would be needed for heating and more time could be spent in pleasant outdoor activities. The projected climate change would, however, threaten the traditional Inuit ways of life. For example, Arctic mammal populations would become too small to support hunting.

Should we try to prevent climate change in this case? According to utilitarianism, we should not. On that view, we should do whatever would satisfy the largest amount of preferences. However, more preferences will be satisfied overall by satisfying some preference of every Canadian, even if this means failing to satisfy the central preferences of the small Inuit population. Therefore, utilitarianism implausibly makes the Inuit pay the price of small improvements to the lives of all Canadians.

According to contractualism, we should instead compare the personal burdens which individuals would need to bear under the alternative policies. The Inuit would have to bear very serious personal burdens if we failed to prevent climate change in this case. They would lose their traditional livelihoods. In contrast, the only personal burden any Canadian would have to bear from climate change mitigation would be the loss of the small benefits gained from the warmer summers and milder winters.

Because of this, in this case, the Inuit are threatened by far more serious personal burdens than the Canadians. The Inuit could use these potential serious unnecessary burdens to reject any policies that would fail to prevent the climate change in question. This shows how contractualism can protect individuals and small groups.

(*d*) The Costs of Mitigation and Adaptation: The final part of the climate change debates which I have room to address is the debate about how the costs of the climate change mitigation and adaptation should be distributed (Singer 2002: ch. 2; Jamieson 2005; Baer 2006).

The 'Mitigation Policy' discussed earlier would severely limit the amount of greenhouse gases that could be emitted. The remaining, reduced emission capacity needs to be distributed to individuals in some way. The more one is required to cut emissions, the more costly it is for one to comply with the relevant climate policy.

Also, even the 'Mitigation Policy' leads to some climate change. There are two ways in which we can diminish the harm caused by the resulting warming and the more frequent severe weather events. We can prepare ourselves for these events in advance as well as use our resources to repair the damages afterwards. Both of these options are costly.

How should these costs be distributed? The main alternatives in the current literature are (*a*) the 'polluters pay' policies, (*b*) the capability policies (that is, the wealthy will pay), (*c*) the egalitarian policies (that is, everyone will pay), and (*d*) various mixed policies. In order to choose between these options, according to contractualists, we should compare the worlds in which these alternative policies have been adopted. We should then consider what kind of lives would individuals live under them. The correct policy would then be the one which would create the least serious personal burdens.

It would be very likely that the non-rejectable cost-distribution policy would include a requirement for the polluters to pay a majority of the costs. This requirement would have beneficial deterring consequences. Given that the polluters would be aware of this publicly enforced policy, it would often be less rational for them to pollute because of the increased costs of polluting. This would mean fewer emissions and less serious climate change and personal burdens.

However, the non-rejectable policy would also include a requirement for the wealthy to pay some of the costs of mitigation and adaptation. Demanding the polluters to pay might make all beneficial economic activity prohibitively expensive, which would be bad for everyone. Thus, some costs would need to be distributed on other grounds.

Demanding everyone to pay would require that the poorest would have to pay some of the costs too. However, they could use the intrapersonally aggregated burdens of having to both (*a*) live in poverty and still (*b*) pay for the costs of mitigation and adaptation to reasonably reject the 'everyone pays equally' policy. The alternative policy that requires both the polluters and the wealthy individuals to pay for these costs would not cause equally serious burdens to anyone. This is why such a mixed policy could probably not be reasonably rejected.

To conclude, in this chapter, I have sketched a contractualist framework for thinking about climate change as an ethical problem. Even if this framework is incomplete in many ways, I hope I have done enough to show that the contractualist perspective deserves to be investigated further and taken more seriously in public debates on climate and climate change.

References

Baer, Paul. 2006. 'Adaptation: Who Pays Whom?', in W. Neil Adger, Jouni Paavola, Saleemul Huq, and M.J. Mace (eds), *Fairness in Adaptation to Climate Change*, MIT Press, Cambridge, 131–53.

Bayles, Michael. 1976. 'Harm to the Unconceived', *Philosophy and Public Affairs* 5(3): 292–304.

Caney, Simon. 2009. 'Climate Change, Human Rights, and Moral Thresholds', in Stephen Humphreys (ed.), *Human Rights and Climate Change*, Cambridge University Press, Cambridge, 66–90.

Gardiner, Stephen. 2006. 'A Perfect Moral Storm: Climate Change, Intergenerational Justice, and the Problem of Moral Corruption', *Environmental Values* 15(3): 397–413.

Hanser, Matthew. 1990. 'Harming Future People', *Philosophy and Public Affairs* 19(1): 47–70.

Intergovernmental Panel on Climate Change (IPCC). 2013. 'Summary for Policymakers', in Thomas F. Stocker, Dahe Qin, Gian-Kasper Plattner, Melinda M.B. Tignor et al. (eds), *Climate Change 2013: The Physical Science Basis*. Contribution of Working Group I to the Fifth Assessment Report of the Intergovernmental Panel on Climate Change, Cambridge University Press, Cambridge.

Jamieson, Dale. 2005. 'Adaptation, Mitigation, and Justice', in Walter Sinnott-Amstrong and Richard Howarth (eds), *Perspectives on Climate Change*, Elsevier, Amsterdam, 221–53.

Parfit, Derek. 1984. *Reasons and Persons,* Oxford University Press, Oxford.

Pettit, Philip. 2000. 'A Consequentialist Perspective on Contractualism', *Theoria,* 66(3): 228–36.

Scanlon, Thomas M. 1998. *What We Owe to Each Other,* Belknap Press, Cambridge.

Singer, Peter. 2002. *One World: The Ethics of Globalization,* Yale University Press, New Haven.

Shue, Henry. 1993. 'Subsistence Emissions and Luxury Emissions', *Law & Policy* 15(1): 39–59.

Suikkanen, Jussi. 2011. 'Contractualism and the Repugnant Conclusion', http://sites.google.com/site/jussisuikkanen/wip (accessed 30 April 2013).

8

Kamma, Virtues and the Individual: An Early Buddhist Perspective on Climate Change

Pragati Sahni

Most scientific research today supports the belief that we are faced with global warming and changing climate patterns. The effects of this warming, though not fully known, are likely to put further stress on a delicately balanced eco-system. Social scientists consider that these effects will lead to secondary problems, such as large-scale migration, poverty and water wars. Human actions are held largely responsible for global warming. If it is believed that global warming and climate change threats are indeed real and that human acts are responsible for them, then inputs from the ethical sphere are warranted.[1] This sphere dictates that notions of human responsibility towards global warming be ascertained through moral debate; prescriptions for correct moral action be found; and a basis for consideration of all affected be established — irrespective of whether these are individuals or communities, developing or developed parts of the world, humans or non-humans, present or future generations.

In addressing these matters, ethical theory is presented with a substantial challenge. To then find an ethical response in an ancient system of thought such as Buddhism, where climate change, let alone its implications, was impossible even to conjure, seems like an especially daunting prospect. Yet I turn to Buddhism. I think of this as an extremely important exercise to re-establish the universality of the ethical truths proposed by the Buddha. There may just be something within Buddhism that rises to the challenge of climate change even though the religion says nothing akin to modern talk on climate change.

---

[1] See Brown et al. (n.d.). One ought to be aware, however, that contrary opinions persist about the ascertaining of a climate crisis. For example, see McKitrick (2005).

To substantiate my belief that Buddhism may indeed contain some responses, I will focus particularly on two questions of considerable importance (both connected with the role of the individual) in climate change ethics: 'Am "I" to blame?' and 'What ought I to do?' The first is about individual responsibility for global warming, and I address this (in the first section of the chapter) through the theory of *kamma* (*karma*). The second question I attend to (in the second section) through Buddhist morality, focusing on some virtues prescribed within it. Before I begin to address these questions, I would like to shed light, albeit very briefly, on the Buddhist world view. I would also like to add that in this chapter, though aware of Buddhism's expansive and varied history, I have limited myself to its earliest written literature as contained in the Pali Canon (and I generally refer to this as early Buddhism).[2]

Buddhism distinguishes itself from other religions, at the very onset, in its understanding of God; gods exist for a time span only. Centrality is accorded instead to the theory of kamma. This theory amounts to saying that beings are responsible for their actions which produce inescapable consequences in this or another life. Each life (godly or any of the lower, unhappy ones, including human) is owed to the nature of actions the being has previously committed, and no life is permanent. Being repeatedly born as a result of actions is not a desired goal: it is full of sorrow or *dukkha*. The notion of dukkha is the basis of the Four Noble Truths which describe the nature of dukkha, its cause, its removal, and the 'path' to remove it. To pursue the latter, Buddhist disciples must be trained in *Sīla* or morality, *Samādhi* or meditation and *Paññā* or insight (these three contain eight constituents that make up the Eight-fold Path). Perfection of training in these signifies *nibbāna* or liberation from constant rebirths in a world full of sorrow.

[2] The Pali Canon has three divisions or *Piṭaka*s namely the *Sutta, Vinaya* (henceforth Vin) and *Abhidhamma*. The *Sutta* Piṭaka is divided into the five *Nikāya*s: *Dīgha, Majjhima, Saṃyutta, Aṅguttara* (henceforth abbreviated as D, M, S, and A, respectively), and *Khuddaka* (the latter containing several texts). In this chapter, I mostly rely upon portions taken from the *Sutta* (discourse-centred) and *Vinaya* (discipline-centred) Piṭaka. I have not included commentarial material. (References to the *Nikāya*s and *Vinaya* henceforth include the volume and page number of the original Pali Text Society [PTS] edition. Verse numbers are mentioned for *Dhammapada* [henceforth Dp]. Translations are included.)

Kamma, Intention and Responsibility

The introductory part raised an important question about the responsibility of human beings for climate change. Many believe that this is a collective and political problem rather than one related to individuals. Collective actions cause global warming, and people place their trust in political institutions and other such bodies representing the collective to make a difference by — say — passing laws to curb greenhouse gas (GHG) emissions. But can the individual agent be completely absolved of all responsibility regarding climate change? Many believe not, and ethical studies have been undertaken to investigate the individual's role. Such enquiries, however, are generally considered quite complex for a number of reasons (explained subsequently). Early Buddhist ethics is about the individual in the sense that its central philosophy revolves around individual moral acts, their consequences and the strife for liberation. In this section I look at the theory of kamma, which is the primary idea that addresses moral responsibility. I shall keenly investigate some aspects of kamma-related ethics to find out if they can prove useful in delineating individual responsibility for climate change.

Walter Sinnott-Armstrong (2010) gives a very good rendering of why it is difficult to pin moral responsibility for climate change on any single individual. He goes through several arguments whereby the driving of a gas guzzling SUV vehicle for 'fun,' even though it comes across as wasteful driving, may not be violating any moral principles. He explores the principle of not harming others and believes it is not violated by this act of driving because first, global warming will happen even if the driver does not take this joy ride; secondly, global warming will not happen unless many others also expel GHGs. So the act of driving the SUV is neither necessary nor sufficient for global warming (ibid. 334). Furthermore, the intention of the driver is clearly not to cause harm. Sinnott-Armstrong takes up the indirect harm principle next. He suggests here that even if the driver's act results in others following his example and tempts him for more drives, his act is still not immoral as he never intended harm and his action has led to no climate change. He claims that '[t]he scale of climate change is just too big for me to cause it, even "with a little help from my friend"' (ibid.: 336). Among internal principles, Sinnott-Armstrong discusses what he terms a virtue principle: 'We have a moral obligation not to perform an act that expresses a vice or is contrary to virtue' (Sinnott-Armstrong 2010: 338). He finds it difficult to point out what virtue would be undermined by driving the SUV or to identify a reason that determines that such driving is indeed an expression of a vice. Sinnott-Armstrong argues that we may focus on green virtues, such as moderation, but this still doesn't provide a specific obligation. How about actions done by a group of individuals? There is a moral obligation not to act in a way that 'makes us the member of a group whose actions together cause harm' (ibid.: 340). Even so, the driver did not intend harm by driving and neither can he change how the group acts. So, by not driving he may be doing a good thing but he is still not morally obligated not to drive. Sinnott-Armstrong, thus, draws attention to how difficult it is to individuate a moral principle that would bind an individual to act against the release of GHGs.

His paper focuses on finding a legitimate basis to enforce individual moral principles or obligations in relation to climate change. Morality for him appears as other-regarding. Though the Buddhist world view and its application of morality are very different I would like to consider a Buddhist response to some of the concerns that have played an important part in the discussion earlier, in the hope that the insights that follow are somewhat valuable across all moral points of view. The concerns I take up are to do with intention and group acts. To investigate intention from the Buddhist point of view (keeping in sight Sinnott-Armstrong's important question whether moral responsibility for climate change can be pinned on an individual who does not 'intend' it), more needs to be said on kamma.

The theory of kamma was ethicized by the Buddhists and in this they adopted a pre-existing concept. The Buddha is said to have described kamma as 'intentional action . . . . Having intended, *kamma* is done by body . . . speech . . . thought' (A III: 415). The definition clearly points out that an action can be regarded as kamma only if it is intentionally willed. Kamma is a process that includes the intention and the actual action and implies that a consequence will follow. Intention becomes important because once corrupted it can go on to influence the nature of the action in like manner leading to *akusala* or unwholesome kamma; it follows that a lack of intention to cause harm, even though harm may follow, suggests that the agent cannot be held responsible. Once intended and acted out, the effect of the action accrues not only to those towards whom it is committed but also to the agent himself. Interestingly, the latter appears to have been given more attention and this adds to the belief that the religion is mostly focused on the individual and the acts he commits. The literature goes on to specify that once kamma has been undertaken by the agent its consequence cannot be avoided: it may be expressed in the agent's lifetime or may determine the nature of his next life (for example, see M III: 203–6). Most importantly, however, kamma shapes the character of the agent. As Dale Wright puts it, 'Karma is the teaching that tells practitioners that it matters what they do throughout their lives, and how they do it. It articulates a close relationship between what one chooses to do and who or what that person becomes over time' (2005: 79). The agent is constantly reminded that he is the cause of his act and he is the heir to what follows (M III: 203). At the same time he is told that he can make a positive difference to both — through performing kamma that is *kusala* or wholesome. In Buddhist morality there is enough scope for the individual to improve his character as well as the nature of action undertaken. The notion of kamma may be taken as the most fitting way in early Buddhism to speak about moral responsibility.

So far morality, expressed through the framework of the theory of kamma, appears to be mainly self-regarding. However, this isn't the complete picture, as can be known from an assiduous study of intention. The *Vinaya* text that deals with disciplinary rules for monks can be studied to understand intention in application. I look at a section within the *Vinaya* that deals with culpability and punishment applicable to monks for killing or harming others in certain situations. First and most importantly, it is generally maintained that when a monk acts in ignorance or without intent to harm he cannot be held responsible for a harmful outcome that includes causing a death. For instance, when a monk had some meat stuck in his throat, in order to help him another monk struck at his throat. The monk died. The offending monk was remorseful but not culpable as he did not intend to cause the death (Vin III: 80). This example reflects the understanding of intention in the Canon. However, the *Vinaya* does suggest that when a risk is taken, even with a harmless intent, the resulting harm can attract some punishment, though not the severest (ibid.: 79). This idea is implied in a story where one monk gave poison to another in order to test it and the latter died. It is recognized that the monk was taking a risk even though he did not intend to cause death, and thereby committed a grave offence. In another case, a monk sat on a boy who was concealed by a rug on a chair causing his death. Death was not intended, but the monk's act is considered punishable, though once again not severely (Vin III: 79). This instance appears to acknowledge the inattention of the monk. In another example, some monks throw a stone for fun down a cliff that kills a cowherd. The Buddha admonishes the monks for their action, which is once again considered an offence, though one of a milder sort (ibid.: 81).[3] It appears from these examples that lack of intention can be cited as a factor due to which an individual can escape culpability and consequences, but in some cases, despite this, the agent must bear at least some responsibility. Punishment follows from undertaking actions with undue risk or inattention or heedlessness that end up harming someone. In these examples, the idea of intention becomes quite complex. It appears here that it is not only the attitude with which the monks' intentions are tempered but also the consequences of their actions on others that is the deciding factor for punishment.[4] This, I believe, is not to question the centrality of intention but to acknowledge other factors that intermittently get involved when an action takes place.

Peter Hershock draws attention to another detail. He believes that 'intentions arise within specific horizons of relevance — that is, within particular meant environments, the precise extent and topography of which are shaped by abiding sets of values. This is often overlooked in expositions of karma stressing its intentional character' (Hershock 2005: 6). In the *Cakkavatti Sīhanāda Sutta*, explains Hershock, the king is confronted with theft related to extreme poverty in his kingdom and decides to help the thief (D; *Sutta* 26). His intention appears to be good — he has sympathy for those suffering and for peace in his realm. However, his action leads to more individuals aspiring to become thieves and

[3] I am not so clear here whether attention is being given to the triviality of the act or its heedlessness.

[4] Intention forms a central part of Peter Harvey's paper (2007) in which he gives a very excellent and detailed Buddhist rendering of unintended harm to the environment. My treatment is more specific, keeping in view the questions I raised in the introduction.

other unexpected outcomes smearing his reign with chaos and violence. Hershock elucidates: 'The king's mistake is not one of intention, but rather of errant commitment to the strategic value of control' (2005: 6). He adds:

> A strictly deterministic understanding of karma focused simply on actions in and of themselves, apart from motivations and evaluative context; is not compatible with directing the dynamics of interdependence in a nirvanic . . . manner . . . one must go beyond experiencing things simply happening as a function of linear and mechanical causal chains, to experiencing things occurring as outcomes/opportunities (ibid.: 6–7).

I agree with Hershock when he points out that the king's intention was correct but there was something lacking in his understanding of the situation and the methods employed. He appears to be saying that isolating actions and intentions without factoring in the setting within which they play out, their underlying motivations, causes and possibilities would yield an incomplete picture.

In another example from these texts, some beings are described as wishing to live without hate and hostility and yet they end up harming each other.[5] So, the intention is correct and yet what follows does not echo this. The Buddha explains that this happens when beings are bound by jealousy and avarice, which arise from like and dislike (and these from desire and so on). Once again the sense here is that the intention is right but the acts that follow are not, because of other perturbing factors present within the individual himself. Thus, far from being straightforward, the idea of intention is acutely complicated. A harmless intention can become infected by the agent's carelessness, risk-taking or inattention. Situations where harm follows a harmless intention must be contextualized to make sense of the conflict. And then a good intention can lead to an opposite action if the motivating roots are corrupted. In all, at one end we have good intention leading to a good outcome and at the other end, harmful intention leading to a harmful outcome. In between these two ends, however, there are shades of grey. The texts, through these examples, appear sensitive to the agent's immediate intention as well as to what underlies the intention. In reducing the punishment for monks, the *Vinaya* is not only acknowledging that the monks did not intend harm: it

---

[5] Hershock has discussed this example as well (see D II: 276f).

is also acknowledging that they cannot get away with their actions completely due to the other factors, as mentioned earlier. The latter is an important part of adjudicating the nature of intention. The examples here are not specifically related to the individual's responsibility for global warming as such, but throw open the question of the morality of the intention itself. A more complete, somewhat parallel, analysis of intention may be in order in climate change ethics to determine the nature of the agent's act.

I now examine two illustrations of the fruition of *kammic* acts that is of some significance in relation to Sinnott-Armstrong's understanding of collective or group acts and the responsibility of individuals acting within the group. Buddhist literature constantly mentions how good and bad consequences follow from good and bad acts respectively. However, barring few exceptions, it leaves the exact relation between acts and their outcomes unspecified. Bhikkhu Payutto says: 'the process of kamma fruition is extremely complex . . . beyond most people's comprehension. In the Pali it is said to be *acinteyya*, beyond the comprehension of the normal thought processes' (1992: 53). Even less is said on acts performed by a group of individuals and their subsequent consequences.

Nevertheless, there are some descriptions of what can be looked upon as group acts. For instance, once the Buddha was asked why the population of humans had decreased so much (A I: 159). He replied that people were burning with lust, wrongful desires and incorrect views. As a result, they were prone to violence and to kill each other and this led to the decrease in population. Interestingly, the Buddha points out in the very next passage that due to such immorality the rains have stopped, crops have suffered, and food is short: all indicating famine and starvation. In another example, the ruler of a kingdom is indicated as unrighteous (A II: 74). This unrighteousness filters down to all in society, including ministers, householders and villagers. The result of this unrighteousness lends itself to constellational disturbances, upsetting patterns of days and nights, seasons, winds and so on. As a result, vegetation is affected and people suffer from sickness and short lives. By cultivating morality, however, all these physical woes go away and the realm is happy and healthy. Morality can make a drought-like situation disappear and the practice of virtues can bring rain.

These two examples, though rare, introduce an important idea that has interested scholars before, referred to as group kamma.

The examples suggest that moral wrongdoings of several individuals can be pooled together to cause catastrophic results. Can the individual be then absolved, for he alone could not have caused the catastrophe or indeed have intended it? The answer is possibly negative. Moral responsibility, both in terms of the action and its consequence, is quite individualistic in Buddhism. The individual naturally gets what's coming to him due to the actions he committed. The scholar James McDermott believes that the only way one can speak of group kamma is through a sense of confluence of the individual kammic reward and punishment of those involved in a given situation (1976: 77). This implies that consequences of many individual kammas fructify at the same time, making it appear that a group of people together has caused something. This understanding clearly does not absolve each agent of the responsibility for his act. But again, is he responsible for the larger consequence?

The Buddhist explanation can be taken to be that rains may stop and a drought may occur when there is general immorality because

> no samsaric stream of existence is completely independent. Although each individual is heir to his deeds alone, the ripening of his karma has consequences that reach beyond himself ... the point maybe simply that in any given situation the karma of each individual must be in confluence with that of every other participant in that situation (ibid.).

Dale Wright too questions this idea of complete isolation of consequences (albeit, from a metaphysical point of view) when he suggests examining

> the extent to which karma can be adequately conceived as a consequence or destiny that is individual, as opposed to one that is social or collective ... Perhaps most strikingly, the view that my acts and their repercussions remain enclosed in a personal continuum that never dissipates into the larger society and continues to be forever 'mine' reinforces a picture of the world as composed of a large number of discreet and isolated souls, a view that a great deal of Buddhist thought has sought to undermine (2005: 85).

Thus, it can be learnt that the possibility that consequences of actions can extend beyond the agent himself in ways one may never comprehend exists quite credibly along with the idea that each agent suffers the consequences of his own act. Responsibility lies clearly with the individual. It can additionally be suggested that in these instances the horror of the larger consequences are pointedly discussed by the Buddha as a pedagogical tool to inspire agents to reflect upon their actions and continually work on improving their character. Clearly, this way of understanding is far removed from Sinnott-Armstrong's analysis, but it suggests that individual responsibility is not to be overlooked, however negligible or far removed from the consequence the act of the individual may seem. It also calls for deep reflection on the act itself.

The Nature of Virtues

A good way to engage with the second question raised in the introduction, 'What ought I to do?' vis-à-vis climate change is by building on the general morality present in early Buddhism. Morality creates dispositions that mostly ensure good intention and right action and reinforce good character. Buddhism is often believed to resemble some version of virtue ethics more closely than any other ethical theory (strands of deontology and consequentialism do occur) precisely because the agent's character is central to judging the rightness of his act. Insight and reflection remain significant too, for along with morality they are required for character-building and spiritual enhancement. Buddhist morality does not directly address climate change, but some moral and other insight-related concepts within it can be looked at as especially relevant to this area. By building on character, the agent can become additionally sensitive to the problems posed by the changing climate scenario. As an illustration, I shall briefly look at Buddhist attitudes to 'right view', humility and contentment.[6]

Early Buddhism speaks much of embracing right view. Right view or *samma diṭṭhi* is a constituent of the Eight-fold Path (under insight) and can be seen to underlie morality. Right view leads to actions that are undertaken with the reflection that they cause no harm and arise neither from greed, hatred nor delusion. It is also suggested in the texts that armed with right view it is not possible

[6] I have earlier written that humility and contentment can be seen as environmental virtues and early Buddhism as an environmental virtue ethics (see Sahni 2008).

that an individual could do something wrong (M III: 64f). Furthermore, the nature of right view includes *recognizing wrong and right intention for what they are* (M III: 71–72, emphasis mine). Peter Harvey sums up the Buddhist position on view succinctly:

> It is said that wrong view leads on to wrong thought, and this to wrong speech and thus wrong action, while right view has the opposite effect. As wrong actions thus come from the misperception of reality, they can be seen to be 'out of tune' with the real nature of things. As they thus 'go against the grain' of reality, they naturally lead to unpleasant results (2000: 17).[7]

It thus appears that by nurturing the right view through awareness, reflection and engagement, the course of actions and the effects that follow can be positively transformed. The possession of the right view related to climate (and environmental) ethics discussions can be useful not only to clarify intention but also to understand what is happening to our planet and to recognize immediate as well as long-term implications of our actions. Inculcating knowledge and awareness about GHG-related situations can be included under right view. The latter could, therefore, be an important part of a Buddhist methodology, which can be set up for addressing many difficult climate-related problems.

Humility or *hirī* appears as a central virtue in Buddhism. This is reinforced through fairly recurring instances of the condemnation of arrogance throughout the literature. In general terms, humility presupposes understanding one's limitations, arrogance or *ahaṃkāra* suggests undue self-importance.[8] It becomes an important trait in climate change ethics as it acts as a reminder of how little we know about the phenomenon and that we may not have all the answers. Arrogance in Buddhist literature is believed to block spiritual growth. The monk must cultivate humility, among other virtues, otherwise his pursuit of wholesomeness and spirituality will be negatively affected (S II: 206). Due to beings not practising humility, in the Buddhist cosmic myth the world disintegrated into something less ethereal and pure (D III: 86f.). The myth seems to

[7] Here, Harvey is referring to A: 211–12 and M III: 66.

[8] Humility is an important virtue in environmental ethics for Thomas E. Hill Jr in his seminal article (2002).

suggest that cultivating arrogance can instigate major changes in the environment. In the current scenario, the myth is reminiscent of our less than humble preoccupation with our needs to the exclusion of all else, along with the dawning realization that our arrogance has indirectly caused much damage to the environment and climate patterns on this planet. Approaching issues with humility and a lack of arrogance may, thus, prove quite valuable to climate change initiatives.

Another value addressed in Buddhist texts is contentment or *santuṭṭha*. I treat contentment here as the opposite of greed — the insatiable desire for more goods and wealth. Greed, along with hatred and delusion, forms the three unwholesome roots of action in Buddhism. By contentment, Buddhism does not propose absolute material non-possession; even renunciants are allowed some material goods including robes and medicine. But they are constantly warned that sense-pleasures and material goods can lead to attachment (see M I: 32). Monks are discouraged from using luxury items, such as high beds, silks, gold, and silver, reconfirming the distracting and ensnaring nature of sensual goods (D, *Sutta* 1). They are asked to be content and contentment is praised as the greatest wealth (A II: 26f. and Dp: 204 respectively). The texts specify:

> Greed ... is unskilful ... One in the power of greed, sunk in greed, whose mind is distorted by greed, causes trouble for others by striking them, imprisoning them, crushing them ... thinking, 'I am powerful, I am mighty' ... These many kinds of coarse, unskillful conditions, arising from greed ... persecute the evil doer (see Payutto 1992: 35).[9]

Greed and arrogance appear linked here. It is warned that greed and wickedness can lead to long-drawn suffering (Dp: 248).

At the same time, even though monks cannot earn or possess beyond the basics, wealth production and possession is recognized as a legitimate pursuit for laypersons. In fact, the possession of wealth is believed to be one outcome of good kamma. Buddhists texts seem in tune with the economic and cultural significance of wealth and in this their approach is practical. However, the production and spending of wealth comes with strict moral restrictions. It appears that Buddhism recognizes the nature of desire and greed

[9] I use Payutto's translation of A I: 201 here.

in relation to wealth; incorrect usage is strongly discouraged.[10] It is said that the agent (*a*) who seeks wealth lawfully, (*b*) makes himself happy, (*c*) uses it for meritorious deeds, (*d*) has no greed and longing, and (*e*) is not blind to his own salvation, is praiseworthy. If he fails to meet these standards he is blameworthy.[11] *Dāna* or sharing of wealth with others in need is also a much-mentioned idea (see A II: 66). In all, attitudes to wealth are bound with notions of generosity, non-greed and moderation. To conclude, it is endless acquisition and excessive freedom to spend personal wealth as encouraged in the current economic scenario that stands questioned herein. In a sense, these values make us question the dictates of an economy that only focuses on material gain; in doing so they attack the very paradigm that has led to the dangerous increase of GHGs. Accompanied by the denouncement of craving for material goods, Buddhist views on contentment and the use of wealth are, thus, important in application to climate ethics.

Conclusion

I started out by asking myself a question that has repeatedly been posed in climate change ethics: how can an individual be held responsible for climate change as (*a*) he did not intend for climate change to happen through his act, (*b*) his action is harmless unless repeated by him or others, and (*c*) even if he chose not to act, climate change would take place nonetheless.

I believe that early Buddhism may have, at least, something to contribute to the questions posed.[12] First, it introduces intention as a moral concept that may sometimes need to be reviewed more dynamically. Second, very importantly, it recognizes that the effects of action, howsoever they may express themselves, are always ultimately traceable to the agent's responsibility. Third, it ensures that the moral transformation of character is not undervalued as

---

[10] See D III: 188 and A II: 249 for more on wealth.

[11] A 1: 81–82 contains this description in detail.

[12] Buddhism would have much more to say, for instance, about political leadership, social and economic frameworks, future generations, and other virtues that I must leave for another time. Contributions can also be sought from later forms of Buddhism, engaged Buddhism and countries where Buddhism is practiced.

a means of individual and holistic enhancement. However, the question that still remains is: can these contributions be taken to imply or suggest the presence of individual moral obligations against climate change of the nature that Sinnott-Armstrong was concerned with? I think ultimately not. But, at the same time, I do not doubt their potential for initiating reflection and leading to a re-thinking of all aspects of individual moral responsibility. I end with Dale Jamieson's insightful words that have echoes of Buddhist thinking:

> We should focus more on character and less on calculating probable outcomes ... [in doing the latter] we systematically neglect the subtle and indirect effects of our actions, and for this reason we see individual action as inefficacious. For social change to occur, it is important that there be people of integrity and character who act on the basis of principles and ideals ... what is important to recognize is the importance and centrality of the virtues in bringing about value change (2012: 197–98).

References

Translated Works

Horner, I. B. (trans.) 1969–70. *Vinaya: The Book of Discipline*, Luzac & Company, London.

Ñāṇamoli and Bhikkhu Bodhi (eds) 1995. *Majjhima Nikāya: The Middle Length Discourses of the Buddha*, Wisdom Publications, Boston.

Rhys Davids, C. A. F. and F. L. Woodard (trans.) 2005. *Saṃyutta Nikāya: The Book of Kindred Sayings*, MBD, Delhi.

Thera, Narada (trans.) 1993. *Dhammapada*: The Corporate Body of the Buddha, Educational Foundation, Taiwan.

Walshe, Maurice (trans.) 1987. *Dīgha Nikāya: The Long Discourses of the Buddha*, Wisdom Publications, Boston.

Woodard, F. L. and E.M. Hare (trans.) 2006. *Aṅguttara Nikāya: The Book of Gradual Sayings*, MBD, Delhi.

Secondary Sources

Brown, Donald A., Nancy Tuana et al. n.d. 'White Paper on the Ethical Dimensions of Climate Change', Rock Ethics Institute, Penn State University, US, http://newdirections.unt.edu/resources/climate_change_white_paper.pdf (accessed 15 March 2012).

Harvey, Peter. 2007. 'Avoiding Unintended Harm to the Environment and the Buddhist Ethic of Intention', *Journal of Buddhist Ethics,* 14: 1–34. http://enlight.lib.ntu.edu.tw/FULLTEXT/JR-AN/an147029.pdf (accessed 12 August 2012).

———. 2000. *An Introduction to Buddhist Ethics,* Cambridge University Press, Cambridge.

Hershock, Peter D. 2005. 'Valuing Karma: A Critical Concept for Orienting Interdependence Toward Personal and Public Good', *JBE* 12: 1–29, http://blogs.dickinson.edu/buddhistethics/files/2011/01/hershock01.pdf (accessed on 3 March 2012).

Hill Jr, Thomas E. 2002. 'Ideals of Human Excellence and Preserving Natural Environments', in David Schmidtz and Elizabeth Willott (eds), *Environmental Ethics: What Really Matters, What Really Works,* Oxford University Press, Oxford, 189–99.

Jamieson, Dale. 2012. 'Ethics, Public Policy and Global Warming', in Allen Thomson and Jeremy Bendik-Keymer (eds), *Ethical Adaptation to Climate Change: Human Virtues of the Future,* MIT Press, Cambridge, 187–202.

McDermott, James Paul. 1976. 'Is There Group Karma in Theravada Buddhism?', *Numen* 23, Fasc. 1: 67–80.

McKitrick, Ross. 2005. 'What is the Hockey Stick Debate About?', APEC Study Group, Australia, http://www.uoguelph.ca/%7Ermckitri/research/McKitrick-hockeystick.pdf (accessed 5 May 2013).

Payutto, Bhikkhu P. A. 1992. *Good, Evil and Beyond Kamma in the Buddha's Teachings,* trans. by Bruce Evans, Buddha Dharma Education Association, http://www.buddhanet.net/pdf_file/good_evil_beyond.pdf (accessed 1 May 2012)

Sahni, Pragati. 2008. *Environmental Ethics in Buddhism: A Virtues Approach,* Routledge, Abingdon.

Sinnott-Armstrong, Walter. 2010. 'It's Not My Fault: Global Warming and Individual Moral Obligations', in Stephen M. Gardiner, Simon Caney, Dale Jamieson, and Henry Shue (eds), *Climate Ethics: Essential Readings,* Oxford University Press, New York, 332–46.

Wright, Dale S. 2005. 'Critical Questions towards a Naturalized Concept of Karma in Buddhism', *Journal of Buddhist Ethics* 12: 77–93, http://blogs.dickinson.edu/buddhistethics/files/2011/01/wright01.pdf (accessed 3 March 2012).

9

Climate Change: Who Does What, Why and How

Marcello Di Paola

There are cases in which the build-up of severally innocuous individual behaviours begets some generally pernicious outcome. Climate change (CC) is one of those cases: it is caused by an increased concentration of greenhouse gases (GHG) in the atmosphere, which is caused, in turn, by the GHG emissions produced by billions of severally innocuous individual actions. No single individual can make a difference to the climate system by eating steaks, taking long showers, using air-conditioning devices, or driving cars. But if no individual can make a difference by doing these (or other) things, then no individual is personally responsible for CC. If individual obligations must have anything to do with personal responsibility, and if no individual is personally responsible for CC, then no individual has an obligation to take action against it. In a famous article, Walter Sinnott-Armstrong articulated this view and drew some important conclusions:

> My fundamental point has been that global warming is such a large problem that it is not individuals who cause it or who need to fix it. Instead, governments need to fix it, and quickly. Finding and implementing a real solution is the task of governments. Environmentalists should focus their efforts on those who are not doing their job rather than on those who take Sunday afternoon drives just for fun (2010: 344).

My personal contribution is neither necessary nor sufficient for CC to obtain, or for it to be 'fixed'. Accordingly, engaging in unilateral, self-starting individual practices such as buying fuel-efficient cars and insulating houses is

> all wonderful, but it does little or nothing to stop global warming and also does not fulfil our real moral obligations, which are to get

governments to do their job to prevent the disaster of excessive global warming. It is better to enjoy your Sunday driving while working to change the law so as to make it illegal for you to enjoy your Sunday driving (ibid.: 344).

I contest these conclusions. In what follows, I shall maintain that individuals can make a difference to the climate system, and thus have an obligation to engage in self-starting anti-CC practices. However, because I share Sinnott-Armstrong's concern for efficacy, I admit that for such obligation to be appropriately fulfilled, these cannot be just any individual practices: rather, they must not be as private as to be hard to co-ordinate interpersonally, while being of significant political import (that is, capable of prompting systemic reform) when co-ordinated. Co-ordination on such practices must not itself require or depend on the coercive intervention of governments, for three reasons. First, from a theoretical point of view, reliance on coercion would make my position virtually indistinguishable from Sinnott-Armstrong's. More practically, coercive intervention will generate monitoring and enforcement costs, absorbing resources that governments could better employ otherwise (for example, research on alternative energy sources, programmes of environmental education). Last, historical record justifies no confidence in constructive governmental engagement against CC. In fact, to be efficacious today, the practices in question must perforce amount to explicit acts of resistance against, and applicable alternatives to, current political, economical and cultural infrastructures that issue into the degradation of the climate system — and governments are not outside but rather at the very crux of such infrastructures.[1]

I will be working under three very general assumptions. First, what matters morally is human well-being. Second, agents ought, on a general moral principle, to protect and promote human well-being. That includes but does not reduce to respecting the 'Harm Principle'. Protecting and promoting human well-being (not the well-being of this or that human, but of humanity at large) is a larger principle than not harming anyone. Accordingly, no harm needs to be inflicted onto anyone in particular for such principle

[1] On the reasons for governmental, and more generally institutional inability and unwillingness to tackle CC, see Gardiner (2011).

to apply. That there is no specifiable victim of my Sunday driving does not mean I am protecting and promoting human well-being by Sunday driving, and the absence of identifiable victims does not dispense me from my obligations to mankind. Third, the spatiotemporal location of human well-being is morally irrelevant.[2] Insofar as the effects of CC will have negative impacts on human well-being, no matter how diffused in space-time, agents ought morally to alleviate CC. According to Sinnott-Armstrong, the agents in question can only be governments, as governmental action can make a difference in both causing and fixing CC. My claim is that individual action, too, can make a difference on both scores; hence, the agents in question are also individuals.[3]

In what follows, I shall avoid all talks of intentions. Nobody intends to cause or contribute to CC by driving a car, as causing or contributing to CC is simply not the point of driving cars — that much is clear. I am interested only in whether individual driving (or suchlike) can make any discernible difference to the climate system.[4] If the severity of individual anti-CC obligations must be commensurate with the measure of personal responsibility individuals have for CC, then the difference that individual driving (or suchlike) makes — if any — will establish the severity of individual obligations against CC. The latter I see as verdicts of the general moral principle that all agents ought to protect and promote human well-being, now applied to the specific case of CC.

In short, I agree with Sinnott-Armstrong that individuals must work to stimulate systemic reform; but I disagree with him on why and how they should do so. Unlike him, I think they should because they can make a difference to the climate system and thus,

---

[2] Against spatial discounting see Singer (1972) and Unger (1996). Against temporal discounting see Parfit (1984), Cowen (1992) and Broome (1994).

[3] Nothing I will say is meant to excuse governments from their anti-CC obligations — to be pursued through diplomacy, research and development, national legislation, carbon taxes, or any other means appropriate. Institutional obligations exist, are severe and ought to be urgently fulfilled: but they are not the subject of this chapter.

[4] This means that I will also not be concerned with blame and blameworthiness. I am only interested in whether an individual can be imputed with having made a difference by doing certain things, not in whether she is to be blamed for doing these things.

in a mediated fashion, impact human well-being significantly — a well-being that every agent ought, on a general moral principle, to protect and promote. I also think that the obligation at issue must be discharged not by addressing, negotiating with, or contrasting governments through institutional channels ('working to change the law' through voting, protests, public discussion and criticism, lobbying, and the like) but by living differently, confronting governments with change. Rather than incremental reform through acts of coercive legislation, I thus recommend self-starting transformations of the ways things are done by individuals in their everyday lives.

What a Difference a Drive Makes

Sinnott-Armstrong does not claim that I am no emitter at all, but only that my GHG emissions are neither necessary nor sufficient for CC to obtain. If that is so, then I cannot be personally responsible for CC. I want to challenge this view. To do that, I first individuate a measure of the lifetime aggregate of my GHG emissions: call this my 'Climate Footprint' (CF). This notion gestures at the possibility of isolating and quantifying my personal input to CC — and that, indeed, can be done with some accuracy. Companies in the emission-offsetting market have developed fairly precise measures of individual CF.[5] These quantifications cannot be absolutely accurate, of course; nonetheless, obtaining reliable estimates is relatively simple. Now, it is obvious that however large my individual CF, from the point of view of the atmosphere it remains microscopic. It is, thus, certain that if my anti-CC obligations have to be commensurate with my CF, they really cannot be too severe. Before talking of obligations, however, it must be shown that my CF can make some difference to the climate system.

Suppose, then, that there exists a global aggregate of GHG emissions so large as to constitute the phenomenon of CC. Suppose, next, that I subtract from it my personal CF: never mind, CC still obtains.[6] Suppose my brother does the same thing, and then

[5] See http://www.atmosfair.de; http://www.climatefriendly.com; http://www.myclimate.com?, all accessed 18 March 2014.

[6] What follows is a variation on the sorites paradox, originally formulated by the ancient philosopher Eulybides.

one by one all the members of my family, my friends, my fellow citizens, and then everyone else in the (history of the) world. At some point, all CFs will have been individually subtracted from the initial aggregate. But if none of them was neither necessary nor sufficient for that aggregate to obtain, then the aggregate should still obtain, even though there no longer are any elements on the scene that can be aggregated. This is, if not logically impossible, at least concretely absurd. On another interpretation, the very absurdity of the above idea would show that CC does not exist at all. This contradicts our best science and is therefore, if not logically impossible, at least factually false. Given his claim that individual CFs are neither necessary nor sufficient to cause CC, Sinnott-Armstrong may be forced to accept one of these two interpretations.

As a way out of this cul-de-sac, one may adopt a third interpretation, according to which at some unspecified historical juncture a given CF will come to make not just some, but all the difference. On this view, there may be some critical point — a threshold — beyond which the subtraction of one specific CF would lead from a situation in which CC obtains, to one in which it obtains no more; and, conversely, the addition of one specific CF would lead from a situation in which CC obtains not, to one in which it does.[7] On this model, individual CFs are jointly necessary but severally not sufficient to cause CC when they are placed below the threshold; while they may be jointly necessary and severally sufficient when they happen to be placed at the threshold — for then they shall push the climate system beyond it. When enough CFs will already have cumulated, a given CF will happen to be crucial. Now, although it is extremely improbable, it is nonetheless not impossible that such CF be mine.[8]

A threshold model, then, has it that the appropriate answer to the question 'Do I make a difference?' is 'I might' (Kagan 2011: 141). One has to accept this answer, or fall back on the unpalatable choice described previously, between a concretely absurd and

[7] Shelly Kagan (2011: 105–41) has recently argued in favour of a similar view, though not in application to CC.

[8] Notice that, unlike 'contributionist' alternatives, the threshold model avoids the logical mistake of thinking that what is part of a cause is also cause of a part. According to the threshold view, my CF, if crucial, does not cause a 'part' of CC, but all of it — in the relevant sense, it makes all the difference.

a factually false conclusion. Now if I am, on a general moral principle, to protect and promote human well-being, given the very vast magnitude of the (negative) difference my CF might make, my obligation is to erase it.[9] I must eliminate any possibility that it be crucial, by offsetting it.[10] But that is not enough. By erasing my CF, I indeed only protect, I do not promote human well-being. Climatically speaking, I just leave things as I found them — but I do not make any positive contribution of my own. So, it may appear that I must do something more than just erasing my CF. That, however, contradicts the idea that my anti-CC obligations must be commensurate with my measure of personal responsibility for it. After all, once I erase my CF (completely), I erase the very possibility of me being responsible for CC. So my suggestion is that, rather than doing more than offsetting my CF, I must offset it in certain ways: ways that also promote human well-being, by stimulating systemic reform. I elaborate on this suggestion in the next two sections.

---

[9] Not to prevent it. On my view, one can produce an indefinitely large CF: that establishes no moral wrong, not even *prima facie*. What does establish a moral wrong is that one does not *erase* that CF (completely) — and, as I note a few lines down in the text, that one does not do so in certain specific ways.

[10] J. Broome (2012: 85–95) arrives at the same conclusion: my anti-CC obligation is that of offsetting my CF. He specifies, however, that individual offsetting is morally obligatory for reasons of justice: my CF harms (spatiotemporally distant) people, and that is unjust. Sinnott-Armstrong would disagree, as he denies that my CF can possibly harm anyone. On my view, the issue does not arise: as I noted in Sec. 1, protecting and promoting human well-being (not the well-being of this or that human, but of humanity at large) is a larger principle than not harming anyone. Note, further, that Broome contrasts obligations of justice with obligations of goodness. He maintains that, from the point of view of (the overall) goodness (I could bring to the world), self-offsetting is of little or no value: from that point of view, I must rather take political action to induce my government to 'do what it should' (ibid.: 81). Broome and Sinnott-Armstrong converge on this conclusion, because they share the assumption that no self-offsetting practices can have direct political import (that is, stimulate systemic reform). I contest that assumption. Undermining it will have the effect of undermining Broome's distinction (ibid.: 73–74) between my anti-CC obligations 'as a private individual' (those of justice) and my anti-CC obligations 'as a citizen' (those of goodness).

A threshold model rescues us from the logical cul-de-sac in which we ended up following Sinnott-Armstrong. From an ecological point of view, it encapsulates the fact that there indeed exist tipping points, beyond which complex and global feedback effects are triggered that may alter the climate system in essential, unpredictable and irreversible ways. From a moral point of view, it has the implication of justifying the ascription of anti-CC obligations directly to individuals, without the need for institutional mediations.

Demandingness and Futility

The next question is what an appropriate fulfilment of such obligations might be. My CF is, from the point of view of the atmosphere, smaller than a neutrino to the eye. Hence, while I do have individual anti-CC obligations, they cannot be too severe. But neither can they be fulfilled through a self-referential palimpsest of purely private practices, such as biking rather than driving and installing a windmill on my roof; for these, though commendable in themselves, are patently futile on a grand scale. This violates Sinnott-Armstrong's legitimate concern for efficacy. What is needed, then, are systemic reforms, and prompting these is the obligation Sinnott-Armstrong ascribes me with. I think he is right as far as that goes. If that is so, self-referential individual action cannot be construed as an appropriate way to discharge my anti-CC obligations. But the way I should prompt systemic reform, according to Sinnott-Armstrong, is by 'working to change the law'. On this view, because governments are the only agents that can 'fix' CC, if individuals are to do anything of moral and political consequence against it, they must address governments, and get them to 'do their job' (2010: 344). My obligation is thus that of reforming climate policies through my electoral choices, protests, public criticism, lobbying, and the like. Fulfilling this obligation could deliver real results; it is not a sanctimoniously quixotic, useless enterprise — like biking rather than driving. In short, I can do a lot more by changing the laws that currently license my CC-inducing behaviours, than by changing the behaviours themselves. And that I should not be changing the behaviours in the absence of governmental edicts is, on Sinnott-Armstrong's view, congruent with my having no personal responsibility for CC.[11]

[11] Clearly, self-offsetting is no obligation for Sinnott-Armstrong.

But let us now take it as established, through the threshold argument discussed, that I can be personally responsible for CC. The fact remains that my individual obligations cannot be too severe, as my CF is quite small, and erasing it is all I need to do in order to eliminate the very possibility that it be crucial. So let us say that, even if there are individual anti-CC obligations, fulfilling them (*a*) must not be over-demanding of the individual's time, efforts and resources; and (*b*) must not be futile on a grand scale. I doubt that, if one takes Sinnott-Armstrong's 'institutionalist' route to systemic reform, these two conditions can be fulfilled in today's globalized political landscape. Accordingly, I doubt that is the route individuals should take.

The 'institutionalist' route counselled by Sinnott-Armstrong will sound familiar to those acquainted with debates on global (socio-economic) justice. In that context, institutionalists often claim that the only demand that justice can make on individuals is that of supporting just, and resist and rectify unjust, domestic institutions. Nagel (2005: 120) calls this a 'political conception', whereby obligations of justice only exist among members of a political community, marked by the existence of a shared set of coercive institutions ruling over such members in their own name. There being no global political community in this sense, the loci of justice can only be domestic institutions. The demands that justice makes on individuals are accordingly restricted.[12]

Implicit in the political conception is the idea that, if enough individuals in each country fulfil their obligations of justice domestically, global justice will eventually emerge compositionally. Obviously, this accepts that great injustices will keep on occurring around the world for a long time (ibid.: 146–47). Unjust institutions exist in many countries, particularly non-democratic ones. It will often be impossible for citizens of these countries to resist and rectify them on their own, due to the political speechlessness their governments may relegate them to. So, in many cases, the cost of avoiding demandingness will be futility; I may maintain or establish justice in my own country most successfully, and yet injustice remains rampant in the world. On a global scale, the efficacy

[12] Nagel specifies that the political conception is 'a very convenient view for those living in rich societies to hold' (2005: 126). Those are, in most cases, the same people who drive gas-guzzlers just for fun.

of the institutionalist route in its 'political' version may be quite limited.

But that is not much of a problem for Nagel. Considerations of efficacy play a very marginal role in his institutionalist picture: the argument is primarily one of principle — a relational argument, tying duties of justice to the sharing of a common (coercive) institutional framework. It is primarily on these grounds that Nagel denies individual duties of global justice — it is not the inefficacy of individual action that crosses out such duties, but rather the absence of a relevant political relation grounding them. Even if it turned out that individual action can be more efficacious than its institutional counterpart, in Nagel's view that would not establish any individual obligations of global justice. Absent a world-government, that is, a source of some globally relevant political relation, there simply can be no such thing.[13]

The case is different with Sinnott-Armstrong. He does insist that the reason why 'fixing' CC is a task for governments, and not individuals, is the (putatively) higher efficacy of governments in doing so (as well as their past efficacy in bringing CC about). Such stress on efficacy is due, I surmise, to some recognition of the particular kind of urgency that characterizes CC. The 'fixing' must be done before the climate system is pushed beyond some critical threshold, and the 'disaster of excessive CC' closes down on us. On the other hand, in normal conditions (absent climate change, that is), the establishment of global justice is an enterprise that is simply open for mankind to accomplish — any time, however far into the future. We have all the time in the world to establish global justice (the sooner the better, of course), but there is some glaring (though nebulous) deadline for 'fixing' CC. This is why institutional efficacy matters more in the case of CC than it does in the case of

---

[13] Institutionalists, Nagel included (2005: 131–32), do often propose efficacy-related arguments in favour of their position. However, such arguments are, in most cases, only secondary, and the point just made in the text holds firm. As L. Murphy (1999: 254) puts it, institutionalists accept a fundamentally dualist view: morality (or beneficence, charity, humanitarian concern — as they might differently call it) is for individuals, justice for institutions. Justice is political not moral. The cut-off is thought to be drastic; Murphy argues such severe discontinuity to be unjustified.

global justice — and thus more to Sinnott-Armstrong than it does to Nagel.[14]

So suppose that the anti-CC obligation Sinnott-Armstrong ascribes me with is one I must fulfil in my own country: the government whose climate policy I must contribute to reform is my own. This seems to be what Sinnott-Armstrong has in mind, and it is not unlike fulfilling my duties of justice only domestically. Now, it is sadly obvious that I might reform my government's climate policy most successfully, and yet average global temperatures still rise. CC does not respect national boundaries; if isolated, my government's action against it will be wholly inefficacious on a grand scale. Sinnott-Armstrong may then argue, somewhat parallel to Nagel, that it is not my task to reform the bad climate policies of governments other than my own, but rather the task of those who live under such governments. Much like Nagel, he would then be confronted by the fact that some of these governments might be non-democratic, systematically relegating their citizens to political speechlessness — also with regard to CC. And the countries these governments run may well be major emitters. China is an obvious example.[15]

Again demandingness would be averted at the cost of futility: my proper discharge of the obligation to get my government to 'do its job' against CC may simply turn out to be wholly inefficacious on a grand scale. But while my society can be just, even if injustice dominates the rest of the world, there is no way my government can 'fix' CC without it being 'fixed' worldwide. And such global inefficacy must be a problem, for Sinnott-Armstrong. He could still insist that I am only obligated to push my government to do its job, part of which is to push other governments to do theirs. But how can I be sure that it will so push? No government has a strategic interest in leading a global anti-CC crusade, nor the necessary leverage to do so — as the Copenhagen 2009 fiasco clearly showed.

[14] Three further, more obvious and interrelated reasons are (*a*) that Sinnott-Armstrong is a consequentialist, while Nagel is not; (*b*) that Sinnott-Armstrong marks no deep distinction between morality and justice; and (*c*) that he does not appeal to relational arguments à la Nagel.

[15] If things go on unchanged, by 2030 China's emissions will constitute more than a quarter of all the emissions produced in the world (Falkner et al. 2010: 257).

And even if my government were to lead that crusade, how can I be sure that it will be efficacious, that is, that all (or at least most) other governments will follow its lead? If my government does not try, or is not efficacious, I should be no more justified in disengaging at that point than if I attempted to extricate myself from my anti-CC obligations just by riding bikes and the like. For if my government's job includes getting other governments to do theirs, and my government does not try or does not manage to do that, then I, as an individual, will not in fact have fulfilled my obligations. My government's inefficacy is my own.[16]

Given a concern for efficacy, my anti-CC obligations should then extend beyond borders. I should not only contribute to reforming my own government's inefficacious climate policies, but also those of other, foreign governments. Perhaps, I should even contribute to the reform (or creation) of efficacious international institutions specifically dedicated to CC. But certainly all of this seems over-demanding of my time, efforts and resources.[17] I will spend my days travelling (producing a larger CF in the process), studying relevant legislation, lobbying institutions, or agitating crowds. My existence will be costly, restless, frustrating, and possibly dangerous (if I was to export my anti-CC enthusiasm to some carbon-intensive authoritarian country, for instance). What is worse, demandingness will not necessarily be rewarded by efficacy; it is rather obvious that, for all I can do, I might still achieve very little, if anything (think of Al Gore, who had the best chance against CC any single individual has ever had and still failed by a long shot). All my efforts may simply turn out to be futile.

---

[16] S. Gardiner puts the matter thus:

According to a traditional view in political thought, social and political institutions are legitimate because, and to the extent that, citizens delegate their own responsibilities and powers to them. On this account, if the attempt to delegate effectively has failed, then the responsibility falls back on the citizens again, either to solve the problem themselves, or else, if this is not possible, to create new institutions to do the job. If they fail to do so, then they are subject to moral criticism for having failed to discharge their original responsibilities (2011: 403).

[17] See Murphy (1999: 281) and Pellegrino (2012: 101–4) for parallel thoughts on the topic of distributive justice.

I conclude that, given a concurrent concern to avert both futility and over-demandingness, my obligation cannot be that of pushing my (or any other) government to do its job against CC in the way in which Sinnott-Armstrong imagines. If, as I think, I nonetheless do have an obligation to prompt systemic reform, I must find ways to do so other than engaging governments through voting, protesting, lobbying, and the like.

What to Do on Sundays

I must personally engage in self-starting anti-CC practices of a specific kind — CF-erasing practices that are not as private as to be hard to co-ordinate interpersonally in the absence of a 'law' (as would be biking rather than driving, or eating vegetables rather than meat), and which will be of significant political import when co-ordinated. Such practices must amount to acts of resistance against, and applicable alternatives to, political, economical and cultural infrastructures that issue into the degradation of a climate system that has so far been congenial to humans. They must be capable of prompting systemic reform, but must not demand more of me than my measure of personal responsibility justifies. So the obligation has two sides: one, creative (individuals must come up with practices appropriate) and the other, operational (they must engage in them). Think, as an example, of the practice of urban and peri-urban gardening as a way to produce food.[18] If co-ordinated at the city level, it may promote more self-reliant economies that could substitute or, at least, significantly complement and to the largest possible extent sabotage current infrastructures of provision, based on oil-fuelled agribusiness and globalized import–export of packaged foods, whose costs in terms of GHG-emissions are notable (see Ch. 11, in this volume). It may also promote a more sustainable approach to land use, not just within but also outside cities. Food production, related land-use change and global food distribution together account for about 20 per cent of global GHG-emissions annually.[19] Were urban and peri-urban

---

[18] By 'gardens' I mean urban private and community gardens, verges, allotments, cloisters, orchards, and peri-urban, small-to-medium, agricultural fields.

[19] This estimate is conservative, as it refers to data for the year 2000 reported in the United Nations Environment Programme (UNEP) (2008: 44).

gardening to reduce that figure by even just one-fifth, by lowering urban appetite for globally-traded food (and there are reliable indications that it could do more than that — see Viljoen 2005; Hester 2006; Worldwatch Institute 2011: ch. 10), it would turn out to be one (non-over demanding) practice through which individuals can appropriately (efficaciously) discharge their obligations against CC: a CF-erasing, self-starting practice that can stimulate systemic reform.[20]

The non-coercive emergence, persistence and spreading of urban and peri-urban gardening may be modelled as follows.[21] The practice, conducted in private and communal gardens, should be co-ordinated at the neighbourhood level, thus involving relatively small groups of people interacting iteratively rather than episodically. In such contexts, the benefits obtained by an engaged individual will be visible to those who are part of his/her interaction network (the network of those the individual has direct contact with, such as family, friends and neighbours), allowing for effective interpersonal signalling. Relatively simple agent-based learning dynamics, particularly imitation, will then stimulate wider convergence on the practice. As the latter spreads, its collectively valuable effects (less polluted urban environments, increased aesthetic delight, promotion of local economy, community building, and the like) will become more prominent. In time, the co-ordination scheme will solidify, thanks to a positive feedback from effects (individual and collective benefits) to cause (urban and peri-urban gardening) — until interpersonal expectations stabilize on participation and the practice becomes something like a social norm.[22] The latter may

---

[20] Further offsetting effects of urban gardening will be directly related to GHG sequestration — see *The Climate-Friendly Gardener*, http://www.ucsusa.org/assets/documents/food_and_agriculture/climate-friendly-gardener.pdf (accessed 12 May 2013).

[21] For a more extensive and analytic discussion of such a model, see Di Paola (2012: ch. 4).

[22] Note that no common interests or shared intentions are presupposed: each individual has her own motivations to garden. Also, such motivations need not be moral; even if engagement in anti-CC gardening is responsive to, and may be justified by, the obligation to protect and promote the well-being of humanity at large, my everyday gardening need not be motivated by an explicit concern for that objective. I will engage in gardening for the

then further expand along the update networks of those involved (the network of those reachable through various means of communication — the web, for instance).[23]

The important point here is that a self-starting, CF-erasing individual practice, when interpersonally co-ordinated, can have immediate political impact. Differently put, one can stimulate systemic reform by acting as a private individual, rather than as a citizen; in fact, the sort of self-starting individual practices we are after, of which urban and peri-urban gardening is but one example, will blur that very distinction.[24]

On the present view, then, political action is crucial indeed, but it takes a different form than Sinnott-Armstrong imagines. I operate (rather than vote, advocate, lobby, or agitate) through a self-starting daily practice that is not over-demanding of my time, efforts and resources, but actually brings me concrete benefits (cheaper food, physical exercise, aesthetic delight, and the like). The practice of gardening, unlike driving less and taking shorter showers, can give rise to non-coercive and yet robust collective schemes that can function as bastions of resistance against, and applicable

---

sake of my own plants, and generally my own well-being (and plausibly that of my family, friends and neighbourhood community) — with positive consequences simply precipitating onto humanity at large.

[23] For a technical treatment of the model of co-ordination expounded, see Skyrms (1996). For the interaction/update network distinction, see Alexander (2007). Regarding imitation and feedback effects, see Hardin (2009). About social norms, see Bicchieri (2006).

[24] When discussing ways to erase one's CF, Broome mainly considers purchasing offsetting credits (2012: 85–96). Assuming the two to be equally effective in erasing one's CF, I believe gardening to be a comparatively superior way to do so. First, being susceptible to interpersonal co-ordination, gardening is an opportunity for individuals to shape and explore new forms of political action. Buying offsetting credits, on the other hand, remains a largely private exercise of limited political potential (as Broome himself recognizes). A second reason, mentioned a few lines down in the text, is that garden practices — unlike buying offsetting credits — also bring concrete benefits to individuals, not just costs. Third, garden practices are methods of environmental education. Fourth, they will involve individuals in a character-moulding cluster of activities that will lead them to develop and exercise important environmental virtues. I have discussed these points in Di Paola (2012).

alternatives to, the global status quo — not through direct confrontation with its institutions, but through everyday operations that will unravel its fabric from within.

Conclusion

In this essay, I have rehearsed some arguments in favour of the idea that individuals have a moral obligation to engage in self-starting anti-CC practices. I have then indicated what sort of practices individuals would have to engage in, if that obligation is to be appropriately discharged. Such practices must erase one's CF; co-ordinating on them should be achievable non-coercively (to that effect, they should allow for effective signalling and learning, and be seen as bringing benefits, not just costs); and, when interpersonally co-ordinated, they must be capable of prompting systemic reform. In all this, they must demand of individuals no more than their measure of personal responsibility justifies.

Individuating appropriate anti-CC practices is crucial; no amount of individual time, efforts and resources will make a difference when channelled through inadequate outlets of engagement. The question for individuals, then, is not whether but how to personally engage against-CC. Apt practice-individuation is part and parcel of the proper fulfilment of one's anti-CC obligations. It is, at the same time, a test for individual creativity and a possible chance to expand one's liberty. I have proposed urban and peri-urban gardening as an example of a practice appropriate. There must be more such practices, and they must be found.

References

Alexander, J. McKenzie. 2007. *The Structural Evolution of Morality*, Cambridge University Press, New York.

Bicchieri, Cristina. 2006. *The Grammar of Society*, Cambridge University Press, Cambridge.

Broome, John. 2012. *Climate Matters*, W. W. Norton and Company, New York.

———. 1994. 'Discounting the Future', *Philosophy & Public Affairs* 23(2): 128–56.

Cowen, Tyler. 1992. 'Consequentialism Implies a Zero Rate of Intergenerational Discount', in Peter Laslett and James Fishkin (eds),

Justice Between Age Groups and Generations, Yale University Press, New Haven, 162–68.

Di Paola, Marcello. 2012. *Giardini globali: una filosofia dell'ambientalismo urbano*, LUISS University Press, Roma.

Falkner, Robert, Hannes Stephen and John Vogler. 2010. 'International Climate Policy after Copenhagen: Towards a Building Blocks Approach', *Global Policy* 1(3): 252–62.

Gardiner, Stephen. 2011. *A Perfect Moral Storm: The Ethical Tragedy of Climate Change*, Oxford University Press, Oxford.

Hardin, Russell. 2009. *David Hume: Moral and Political Theorist*, Oxford University Press, Oxford.

Hester, Randolph T. 2006. *Design for Ecological Democracy*, MIT Press, Cambridge.

Kagan, Shelly. 2011. 'Do I Make A Difference?', *Philosophy & Public Affairs* 39(2): 105–41.

Murphy, Liam. 1999. 'Institutions and the Demands of Justice', *Philosophy & Public Affairs* 27(4): 251–91.

Nagel, Thomas. 2005. 'The Problem of Global Justice', *Philosophy & Public Affairs* 33(2): 113–47.

Parfit, Derek. 1984. *Reasons and Persons*, Oxford University Press, Oxford.

Pellegrino, Gianfranco. 2012. 'Beneficence, Justice, and Demandingness: A Criticism of the Main Mitigation Strategies', in Sebastiano Maffettone and Aakash Singh Rathore (eds), *Global Justice: Critical Perspectives*, Routledge, New Delhi, 91–120.

Singer, Peter. 1972. 'Famine, Affluence, and Morality', *Philosophy & Public Affairs* 1(1): 229–43.

Sinnott-Armstrong, Walter. 2010. 'It's Not My Fault: Global Warming and Individual Moral Obligations', in Dale Jamieson, Henry Shue, Stephen Gardiner, and Simon Caney (eds), *Climate Ethics: Essential Readings*, Oxford University Press, Oxford, 332–47.

Skyrms, Brian. 1996. *Evolution of the Social Contract*, Cambridge University Press, Cambridge.

United Nations Environment Programme (UNEP). 2008. *Kick the Habit: A UN Guide to Climate Neutrality*, Progress Press, Malta.

Unger, Peter. 1996. *Living High, Letting Die: Our Illusion of Innocence*, Oxford University Press, New York.

Viljoen, Andre (ed). 2005. *Continuous Productive Urban Landscapes*, Architectural Press, Oxford.

Worldwatch Institute. 2011. *State of the World 2011: Innovations that Nourish the Planet*, W. W. Norton and Company, New York.

10

Climate Change and the Intuition of Neutrality

Francesco Orsi

Some of the current and future consequences of climate change on the human population are relatively easy to put into some sort of ethical perspective. The catastrophic events related to increasing global temperature (will) cause innocent deaths, diseases, homelessness, and food shortage. In order to prevent such consequences, many people are or will be forced to move to safer lands. Either way, their lives are dramatically affected for the worse. Moreover, most of those who can be said to causally contribute to climate change and its short- and long-term bad consequences (roughly, the industrialized world) seem to stand to benefit from such contribution (they are better off for it), whereas most of those who now or tomorrow will suffer its consequences are only worse off for it. Such facts require a more complex ethical analysis, but still one that employs familiar notions of foreseeable harm, reparation and intra- and inter-generational justice.

Other facts relating to climate change, however, require far more speculative ethical reflections. In his writings on population ethics and climate change, John Broome has urged us to face directly the fact that both global warming and the measures that might be taken to reduce its impact will, in all probability, reduce the size of timeless human population: '[t]he timeless population includes Julius Caesar, me, and all the people who are yet to be born' (2005: 404).[1] The deaths mentioned here will naturally remove from the timeless population all the descendants of the dead person. The migrations alluded to will likewise affect the size of the population. And so on. Therefore, it seems that climate change will contribute to the absence of countless people who would otherwise have

[1] See Broome (2004: chs 10–12) for a fuller account. For the purposes of this chapter, I will discuss Broome (2005).

existed. Now, if this is an additional bad consequence of climate change, then such badness, given the enormous numbers involved, threatens to 'swamp' the badness of the killings of actual people.

Such 'swamping' can mean a number of things. First, in the unlikely event that climate change will in fact add members to the timeless population, and such addition is to some extent a good thing, then such goodness will outweigh the badness of the deaths actually attributable to climate change. This seems to be rather counterintuitive. Second, more realistically, if the measures taken to reduce the impact of climate change will also reduce the size of the timeless population, then this is a bad consequence that must be weighed against the goodness of saving many actual lives. Given the enormous numbers of people who would otherwise have existed, reducing the impact of climate change might turn out to be overall bad. This, again, is not something we are inclined to accept, unless perhaps in the remote event that reducing the impact of climate change, while saving lives, will still lead to a gradual population collapse up to the point of human extinction (discussed subsequently).

Maybe these scenarios are too unlikely to be seriously considered, and the badness of reducing timeless population will simply add to the badness of actual deaths, so that there will be even more good reasons to fight against climate change. Still, the former bad consequence will 'swamp' the latter: if bad at all, then the absence of a potential infinity of humans that would otherwise exist might be worse than even the millions of deaths of actual people. This consequent still seems intuitively false. Intuitively, we ought to care more about, and focus our efforts more on, preventing the actual million deaths brought about by climate change, than on preventing the infinite absences. Preventing actual deaths will predictably also affect the size of timeless population, but this seems, at most, like a fortunate side-effect of our efforts. By contrast, if the reduction in size of timeless population were bad, then preventing actual deaths would mainly matter as an effective means to avoiding what is worst. Putting things like this seems like getting them the wrong way around.

In the following sections, I critically examine the so-called intuition of neutrality as discussed by Broome. In the light of his objections, I urge an alternative normative interpretation of neutrality in terms of an exclusionary permission to disregard the value of adding lives, and argue that it is an intuitive and plausible option. Subsequently, I explore the justification and the limits of such a permission, showing how it deals with the prospect of human extinction. The last section clarifies how the exclusionary permission dispels the 'swamping threat' to our assessment of climate change.

The Evaluative Intuition of Neutrality: Formulations, Objections, Diagnoses

What seems to lie behind our caring more about actual deaths (than about the timeless population size) is a consequence of a common sense view that Broome calls the intuition of neutrality. The subtractions to timeless population that climate change (and the fight against it) will produce — the absence of people who would otherwise exist — simply don't seem to matter: they are evaluatively neutral, neither good nor bad. The same goes for potential 'additions': fighting climate change may also add people to the timeless population, but this doesn't seem like a consequence worth caring about, nor does it make saving actual lives additionally better. In Broome's words: 'Our intuition is that the size of the population is ethically neutral, because we think adding people to the population, or subtracting people from it, is neutral' (2005: 404). The intuitive neutrality of size changes in population seems to follow from the intuitive neutrality of adding or subtracting individuals. According to Broome's: 'A world that contains an extra person is neither better nor worse than a world that does not contain her but is the same in other respects' (ibid.: 401).

Of course we recognize that a new person can make the world better if, for example, it brings happiness to her parents, or can make it worse, by making new demands on our planet's finite resources. The 'same in other respects' phrase is meant to equalize these indirect effects of a new person's existence. But what about the intrinsic goodness or badness of the extra life? The intuition of neutrality seems to hold, quite regardless of whether the extra person will be happy or miserable. Of course, when faced with the possibility of a terrible extra life, most will agree that adding that person's life will make that world worse; and possibly, adding an extremely happy life will make that world better. However, it seems that for a vast range of levels of well-being the existence of the extra person is neither good nor bad: it is good to make people happy, but making happy people is rather different (see Narveson 1973). So here is Broome's qualified statement of the intuition: 'Adding a person whose wellbeing is in the neutral range is neither better nor worse than not adding her' (2005: 406).

Setting the boundaries of the 'neutral range' of well-being is, of course, controversial. But this is not the main problem. Broome's objective is to show that, whatever the inherent difficulties in the neutrality intuition, any attempt to further specify the notion of neutrality in evaluative terms determines the failure of the intuition. This is serious; the stakes are high, as Broome notes: 'only this intuition allows us to be confident even that global warming is a bad thing' (ibid.: 405). So, let us consider two worlds A and B (see Figure 10.1).

Figure 10.1

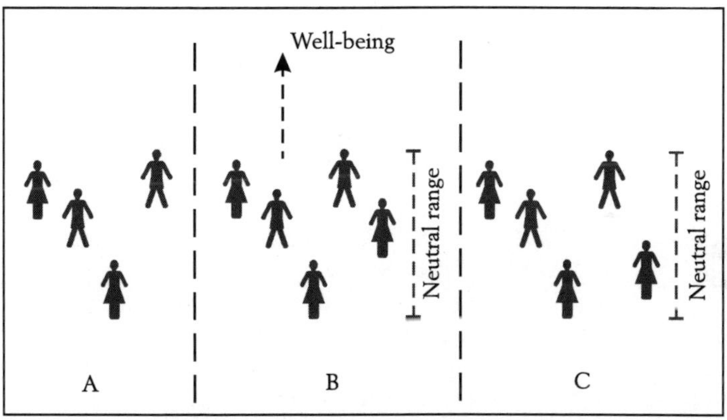

Source: Broome (2005).

A and B are equal in every respect, except that B contains an extra person within the neutral range of well-being. One reading of the intuition says that if A is neither better nor worse than B, then A and B are equally good. Adding the extra person is equally as good as not adding her. Now consider a third world C, in which the extra person is at a slightly lower level of well-being than in B, but still within the neutral range. By neutrality, A and C are also equally good. Given the transitivity of 'equally as good as', it follows that B and C are equally good. But this is unintuitive: since the extra person in C is at a lower level of well-being than in B,

B is better than C. So, an apparently false conclusion follows from interpreting neutrality as equality of goodness.

Another option is to read neutrality as incommensurability in value. Since incommensurability is not a transitive relation, in Figure 10.1 one can consistently say that B is neither better nor worse than A, A is neither better nor worse than C, and that B is better than C. Broome's main criticism against this interpretation is that neutrality as incommensurability is objectionably 'greedy'. The argument starts from considering three further worlds A, B and C (see Figure 10.2).

Figure 10.2

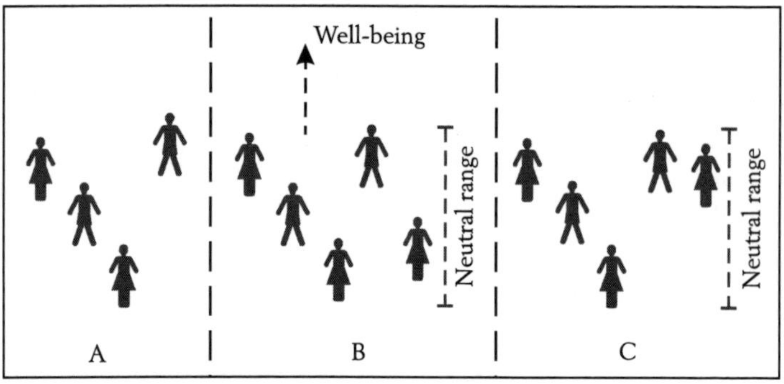

Source: Broome (2005).

A has four people within the neutral range of well-being. B has five people within the neutral range. C has the same five people as B, but a man is worse off than in B (and than in A) and a woman is better off than in B, though both are still within the neutral range. Now, given neutrality, B is not worse than A (Situation 1 represents B and A as incommensurate). Suppose that C is better than B: for example, the woman's well-being is increased more than the man's well-being is decreased (Situation 2: C lies above B). It follows that C cannot be worse than A (Situation 3: C must lie above A) (see Figure 10.3).

Figure 10.3

1	2	3
BA	C B	C (B)A

Source: Prepared by the author.

The problem is that if adding an extra person within the neutral range is neutral, then C does seem to be worse than A. C contains one extra person within the neutral range: that's neutral. But C also contains a man who is worse off than in A: that's bad, or at least makes C worse than A, as far as that goes. A neutral thing plus a bad thing should make C worse than A. But in Figure 10.2, we have concluded that C could not be worse than A.

Now, one way to put the point is simply that the neutrality intuition generates contradictory claims. Broome's greediness objection is, however, slightly more complicated. By generating the claim that C is not worse than A, the neutrality of adding an extra person in C seems to have 'swallowed up' or neutralized the badness of the man's being worse off in C than in A. Likewise, A seems intuitively to be better than C: the extra person is subtracted (neutral), the man is better off than in C (good). But we have already concluded that C cannot be worse than A. So the neutrality of subtracting the extra person 'swallows up' the goodness of the man's being better off in A than in C. This neutralizing effect is what Broome calls greediness. And this is odd: the neutrality of a feature should leave things evaluatively the way they would be, were the neutral feature absent. If C would be worse than A, were the extra person absent, then the presence of the extra person, if neutral, should not overturn that assessment.[2]

To finally fix the nail in the coffin, Broome points out that 'if neutrality is greedy, it cannot do the work we need from it' (2005: 409). Consider global warming. The neutrality intuition is supposed to support the claim that alterations in the size of the population are not what matters in assessing climate change, and our

[2] According to W. Rabinowicz, this idea expresses 'strong neutrality', and as such it should not be greedy by definition. But neutrality as incommensurateness is not a form of strong neutrality, therefore there is no reason why it should not be greedy (2009: 398–99, and fn. 15). In a reply, Broome agrees on the distinction, but still finds the implications of greediness (such that the goodness of saving lives might be swallowed up by the effects on population numbers) 'incredible' (2009: 414). I take him to mean, correctly, that neutrality as incommensurateness (even if conceptually coherent) does not do the justificatory work that the intuition of neutrality was supposed to do (the point is explained in the subsequent section in the text).

response to it. Relatively well-off existences will be 'subtracted' both by climate change and our response to it: but as long as such existences would lie within the neutral range, this fact should not make climate change additionally bad, nor should it make trying to fight against it bad either. When looking only at population numbers, what matters is, principally, the innocent deaths that climate change will cause and which we can prevent. However, it is, in principle, possible that the badness of deaths might be neutralized by the neutrality of changes in the population size, in the way described earlier. Given sufficient neutral population changes, and a sufficiently wide neutral range, a world plagued by the disastrous consequences of climate change might still not be worse than a world without it. As Broome says:

> The change in population caused by global warming will probably be large, whichever direction it goes in. Therefore, if this change is neutral, I think we have to expect its neutrality to swallow up the bad effects of global warming. We shall be forced to conclude that global warming is probably not bad, but neutral (2005: 410).

This conclusion seems both unacceptable and paradoxical: by their neutralizing effect, those very facts which were supposed not to count at all might, in principle, dominate our evaluative assessment of climate change!

Broome's first diagnosis is that the problematic feature in the neutrality intuition is the idea of a neutral range, as opposed to a single neutral level of well-being. In the equality interpretation, the problem stemmed from supposing that extra persons, occupying different positions within the range, did not make the world any better or worse. In the incommensurability interpretation, the problems also stemmed from supposing that somebody could be made better off (or worse off) and still lie within the neutral range. By contrast, if there is a single neutral level of well-being, then adding a person whose life lies above that level will make the world better. And changes in well-being will correspond to ways to make the world better or worse. However, the idea of a neutral range seemed crucial to the neutrality intuition: that a new person would be happy doesn't make adding that person good. Barring extremes in happiness and misery, there is a wide range of levels of well-being that we can simply ignore when considering potential new additions to the human population.

There seems to be a dilemma here. If neutrality is spelled out in terms of a neutral range, then the difficulties pointed out by Broome seem invincible. But if neutrality is reduced to applying to a single neutral level, then it does not reflect a shared intuition, nor does it do its job; for any prospected change in population size, we need to ascertain whether the lives added or subtracted would be above or below the single neutral level. We actually need to count them in our deliberation. Such considerations, given the enormous numbers involved, would threaten to dominate over the tangible effects of climate change on actual existing people. On the original version of the intuition, we could safely assume that population changes would lie within the range, and thus ignore such population changes.

A second diagnosis is hinted at by Broome, when he underlines that the difficulties apply to an evaluative reading of the intuition, where 'neutral' means 'neither good nor bad'. He leaves it open that a normative or deontic interpretation of the intuition might do better (Broome 2005: 412–13). Indeed, the way out of the dilemma that I propose in the next sections takes its lead from Broome's remarks. To develop a normative interpretation means to accept that common sense is confused by its own lights when affirming neutrality as an evaluative intuition: remember that the evaluative intuition was shown to have counterintuitive consequences.[3] But a normative interpretation cannot simply be a matter of stipulation: rather, it must possess an intuitive force comparable to the original, while avoiding its patent problems. *Inter alia,* this means that a normative interpretation will need to make sense of a normatively neutral range of well-being as opposed to an evaluatively neutral range, which was shown to be untenable.

---

[3] Perhaps, a further psychological diagnosis might help to make a normative interpretation more palatable. The intuition of neutrality strikes us as plausible when stated in abstract evaluative terms. But as soon as we try to draw out its implications by picturing worlds against one another, the very presence of a new added person with a certain level of well-being inclines us to perceive an evaluative difference between the worlds, and one which is not naturally captured by incommensurateness (even if other things are equal, there must have been a change, for the better or worse!). If so, evaluative neutrality (either as equality or as incommensurateness) can hardly be said to be intuitive.

Any alternative interpretation of neutrality, thus, will have to meet at least the following conditions:

(*a*) it must be coherent and not 'greedy' (unlike the evaluative reading),
(*b*) it must have an intuitive force comparable to the evaluative reading,
(*c*) it must provide a flexible approach to assessing population changes brought about by climate change (this will matter when considering the possibility of human extinction).

A Normative Model for Neutrality: Exclusionary Permissions

Broome already hints at a particular normative reading of neutrality:

> Think about a couple who might have a child. Our intuition is that their having a child is neither better nor worse than their not having one. But we now know this intuition is mistaken except in the special case where the child happens to live at exactly the single neutral level. So if the couple have a child, that will generally be either better or worse than their not having one. Suppose it is better. Then the couple are in a position to make the world better by having a child. But even so, we might think they have no moral responsibility to do so. We might think they are doing nothing wrong if they choose not to. This normative conclusion about rightness and wrongness may be part of what the neutrality intuition is pointing to. Possibly the intuition might be given a coherent interpretation in these normative terms. And possibly it may apply to grand issues such as global warming as well as to a couple's decision about a child. Global warming will be very good or very bad because of its effect on population. But possibly we may have no moral responsibility towards population, and we may be entitled to ignore the goodness or badness of this effect (2005: 413).

As the last sentence makes clear, a normative reading will put us in a position to claim both that certain outcomes, like changes in the population size, are good or bad, rather than evaluatively neutral, and that it is permissible for an agent to ignore such goodness or badness. The idea may at first sight look paradoxical: how can what is good or bad be permissibly ignored? After all, one might say, if certain things are good then they ought to be promoted or favoured in some way, and if certain things are bad then they ought to be minimized or disfavoured. Ignoring either seems to be impermissible. However, we do have a model in the theory of reasons which can dissolve the air of paradox: Joseph Raz's exclusionary reasons.

An exclusionary reason is a reason to disregard certain reasons and not act on them. So it is a kind of second-order reason. As a good exclusionary reason it will truly justify the agent in ignoring certain first-order reasons. As such, it does not override first-order reasons, nor does it undermine their status as reasons: it simply justifies excluding them from our deliberation. Raz's own examples of exclusionary considerations include promises to act only on certain reasons, orders of an authority (indeed, for Raz, norms in general), and conditions such as fatigue, where an agent acknowledges the good reasons for and against a certain decision but, given her temporarily unreliable mental state, prefers — justifiably — to give up the task of balancing them and coming to a decision (1999: 37–39).

The normative interpretation of neutrality requires the notion of an exclusionary permission: a consideration that does not mandate, but permits ignoring the reasons stemming from the intrinsic goodness or badness of adding new lives within a certain level of well-being. As Raz says: 'Exclusionary permissions differ from exclusionary reasons in that they do not entail that one ought to disregard the excluded reasons. They merely entitle one to do so' (ibid.: 90). Raz's working example is the analysis of supererogation. Even if donating money to Oxfam would be better than spending it for our family, and thus what we ought to do on the balance of reasons, we are permitted to disregard or exclude such reasons and not act on them. An act is supererogatory when we ought to do it on the balance of (first-order) reasons, but we are permitted not to act on the balance of (first-order) reasons (ibid.: 94).[4]

So, an exclusionary permission to ignore the goodness or badness of adding lives would entitle us to disregard changes in population size brought about by climate change and by the policies enacted to reduce its impact. How does the exclusionary model meet the three requirements mentioned earlier? First, exclusionary

[4] Supererogation is only Raz's example: I am not suggesting that taking into account changes in population size is supererogatory, although we could imagine situations where it would be.

permissions are not 'greedy'. Consider again the three worlds A, B and C (see Figure 10.2). Remember that on the evaluative interpretation of neutrality as incommensurability, we were forced to conclude that C is not worse than A, where this was contradicted by the intuition that C is overall worse than A (because C contains one bad feature and a neutral one). For Broome, the evaluative neutrality of adding a person in C seemed to swallow up the badness of a person being worse off in C than in A. This is both implausible on its own right, and a strange upshot of neutrality; the neutrality of a feature should keep things the way they would be, were the neutral feature absent.

On the exclusionary model, the badness of a person being worse off in C than in A — the reason to prefer A to C — is not swallowed up by the fact that we have a permission to disregard the presence of an added person in C, provided that her level of well-being lies within the range in which the exclusionary permission is applicable, what we can call the normatively neutral range. What we are permitted to exclude is the value of the added life, not the change for the worse of the one person existing at both worlds. The point is that C might be overall worse, or better than, or as good as A, depending on the well-being of the added person; in any such case, we are permitted to disregard such value differences as determined by the addition of a new person within the relevant range. The exclusionary model indeed does not even depend on the possibility of comparing the overall value of A and C.

Looking at the issue of climate change, on the exclusionary account there is no danger that the goodness of effective climate change policies or the badness of climate change will be swallowed up by the neutrality of their effects on population size. Neutrality here means precisely that we are permitted for the most part to disregard such effects, and such a permission cannot swallow up positive reasons to fight climate change.

Second, the exclusionary account seems to retain much of the force of the original intuition. The central (and problematic) feature of the evaluative interpretation was the notion of an evaluatively neutral range of well-being levels. Broome's arguments show that we need to abandon such a notion: at most, there will be one single neutral level of well-being, if any sense can be made at all of a life that is neither worthy nor unworthy of living (for example, contrary to hedonistic views, a life devoid of good and bad experiences seems to fall below that level). But the idea of a neutral range as opposed to a single neutral level can and should be kept. The exclusionary permission will entitle to disregard a vast range of well-being levels as reasons to add or to refrain from adding new lives. Such a 'permission range' will have rough limits; intuitively, we would not be permitted to ignore lives that are exceptionally good or exceptionally bad. We have some non-excludable (but still defeasible) reason to create the former, and some non-excludable (but still defeasible) reason to prevent the latter.

Perhaps some might think that the 'permission range' should not include positively bad lives: even if we might be overall permitted to create bad lives in some special circumstances, surely we are not permitted to disregard altogether their badness: we have a non-excludable reason to give it some weight in our deliberation. On the other hand, it seems that reasons to create (less than exceptionally) good lives can be excluded altogether. Now, this is a matter of substantive moral debate that need not concern us here. The exclusionary account simply provides a coherent framework to express our intuitions — if intuitions suggest a moral asymmetry between creating good and bad lives here, the 'permission range' will be defined accordingly. Whether such asymmetry is defensible requires an investigation that falls outside the scope of the chapter.[5]

Returning to practical issues, since we are dealing with a permission here, we are still entitled to take account of changes in population size, and we would still have good reasons to do so. This is in the spirit of the original intuition of neutrality: the idea was not that it would be wrong to consider changes in population size, or that adding people within the neutral range must not matter, but simply that it does not matter. To be sure, the original notion of new lives 'not mattering' suggests the absence of reasons to care about adding them, whereas the exclusionary model acknowledges

[5] A deontological approach might justify the asymmetry by distinguishing strict requirements of non-maleficence (do not create evil) and looser reasons of beneficence. But so can rule-consequentialism distinguish permissions not to create good lives (justified by various benefits, explained subsequently) and requirements not to create bad ones (justified by the obvious, immediate badness of such acts).

the presence of (first-order) reasons, albeit ones that are permissibly excluded. I think this residual tension with the original intuition is a price worth paying for defenders of the intuition, in the light of the difficulties of the evaluative version, and the implausibility of rejecting neutrality altogether.

The third condition requires the exclusionary account to be flexible enough to deal with all sorts of population changes. Is the fact that a new good (or bad) life will be added or subtracted always excludable as a reason? Are potential parents among the few remaining members of a certain ethnic group permitted to ignore the fact that they might create new lives, thus perpetuating the existence of the group? Can the few remaining members of the human species ignore the possibility of perpetuating the species?[6]

Exclusionary Permissions: Grounds, Scope and Human Extinction

To approach this kind of an issue we need to say more about the structure of exclusionary permissions. Two points need emphasizing — an exclusionary permission is based on some ground, and it has a limited scope, that is, it entitles one to disregard some but not all reasons for or against a certain action (Raz 1999: 40, 91): 'the scope of an exclusionary reason [or permission] is the class of reasons it excludes' (ibid.: 46). On the topic of grounds, what could justify a general permission to disregard reasons stemming from the possibility of creating good lives (and bad ones, if we deny the asymmetry discussed earlier)?

Different approaches to moral theory might provide different answers here. For instance, on a person-affecting view one could argue that such a permission is justified in virtue of the unique position of power or privilege that living, actual agents enjoy with respect to the non-existent, potential, future generations. Even if such potential lives would be good for their owners, and consequently there might be reasons to make them actual, still there is nobody actual to whom we owe it that such lives be created. Hence, this is a permission to disregard goodness-based reasons to create

[6] See Broome: 'If the intuition of neutrality is correct, it tells us that extinction is neither good nor bad, provided the future people who will exist if humanity does not become extinct, live within the neutral range' (2010: 112). I argue that once interpreted as an exclusionary permission, neutrality does not entail anything regarding the value of extinction.

them. Notice that this does not imply absolute moral discretion with regard to future generations. For example, we may still owe it to them that, if we create them, then we make sure that their lives will be good. Alternatively, one could argue in the spirit of Samuel Scheffler's agent-centred prerogatives that it is the personal perspective of each of us that grants us such a permission to disregard altogether at least certain ways to make the world impersonally good. Finally, on a consequentialist basis it seems that a permission to disregard potential good lives is justified insofar as everyone having the freedom to pass up such opportunities to make the world impersonally good maximizes overall value. One of the values that such a permission, generally internalized, might (and does seem to) maximize is precisely the enhanced quality of life of those who are actually born, for no other reason than that living in a less crowded family (or a less crowded world) normally is an advantage. If we have reason to create good lives, then overall conformity to such a reason might well be made more likely by a generalized permission not to create good lives whenever one can.[7]

Some justifications might be theoretically more or less plausible than others, but, in fact, it does not ultimately matter which one is the best. The crucial question is whether they allow a permission that is flexible enough to respond to some extreme scenarios in the intuitively right way. The scenarios are those in which the very existence of a certain group or of the human species itself is endangered. In these cases, it seems counterintuitive to suggest that the existence of future people is a possibility we are permitted to disregard altogether. However, here considerations of scope help us to see that the exclusionary account is not committed to the permissibility of letting humanity go extinct. The scope, or class of reasons that are excluded, specifically have to do with the goodness of potential lives for the people that would be added. Insofar as such considerations of goodness are concerned, we are allowed to disregard them and to choose, not to create new lives on that basis. But other reasons to create new lives fall outside the scope of the exclusionary permission, and, thus, may not be permissibly ignored.

This is the case in the extinction scenario. Here what matters is not the goodness of future lives, but rather the possibility there

[7] See Mulgan (2006) for a rule-consequentialist account.

might not be future human lives at all. Intuitively, we are required to take this possibility at least as a *pro tanto* reason to create enough new human lives, and to make sure, now, that such scenarios will not occur in the future. So the exclusionary permission simply does not apply in such cases, and does not conflict with the requirement.[8] Of course, the requirement might conflict with other considerations, such as the relative loss of freedom that policies aimed at preventing extinction might cause.

All three justifications sketched here leave room for the requirement not to let humanity go extinct. The person-affecting approach might claim that we owe it at least to our contemporaries to make sure that their lives, efforts, achievements etc. will be possibly remembered, appreciated etc. and not fall into the oblivion of human extinction. Possibly, we also owe it to the future generations that if we create them, then they can hope to be remembered, appreciated etc. (while still not owing them that they exist). The agent-centred prerogative approach likewise might focus on our personal interest in preventing human extinction; in any case, the scope of the prerogative is itself limited, and should not justify ignoring disasters such as human extinction, even supposing that such disasters are only impersonally bad. Finally, if human extinction implies the irreversible extinction of the greatest values (among which the very possibility of appreciating values), then consequentialism, too, gives us obvious reasons to prevent such a disaster and perpetuate the species.

The upshot of this section is that the exclusionary account of neutrality, given the limited scope of the permission, does not entitle us to disregard the extinction of humanity. Moreover, it is compatible with a number of theoretical justifications, which leave room for a requirement to prevent human extinction. Thus, their plausibility as approaches to questions of population ethics is somewhat enhanced. Since the exclusionary account can receive support from any of them, the account itself comes out as stronger as well.[9]

---

[8] The relevant (positive) consideration here is 'a new human being might exist'. This consideration is, barring religious views, normatively irrelevant in itself: something that we do not need a specific permission to exclude.

[9] I take it that it is not inconsistent to discard the evaluative intuition of neutrality while reintroducing the notion of value in providing the

Conclusion

The exclusionary account seems to provide a viable alternative to the evaluative intuition of neutrality. Remember that the intuition of neutrality protected us from three 'swamping' threats: (*a*) in the unlikely event that climate change will, in fact, add members to the timeless population, and such addition is to some extent a good thing, then such goodness can outweigh the badness of the deaths actually attributable to climate change. (*b*) If the measures taken to reduce the impact of climate change will also reduce the size of the timeless population, then this is a bad consequence that must be weighed against the goodness of saving many actual lives. Given the enormous numbers of people who would otherwise have existed, reducing the impact of climate change might turn out to be overall bad. (*c*) If bad at all, then the absence of a potential infinity of humans that would otherwise exist might be worse than even the millions of deaths of actual people. This consequence seems intuitively false. Intuitively, we ought to care more about, and focus our efforts more on, preventing the actual million deaths brought about by climate change, than on preventing the infinite absences.

The exclusionary account of neutrality leaves the value of changes in timeless population intact. To this extent, the swamping threat is still present. However, a permission to exclude such value and the relative reasons completely silence the swamping threat from a normative point of view. Of course, being a permission and not a requirement, humanity may still, in principle, choose to prioritize changes in population size when deliberating about climate change. But that is insufficiently likely to be seriously considered a shortcoming of the exclusionary approach. Nor is it plausible to suggest an exclusionary requirement here: at the individual level, this would seem to limit too much one's freedom in reproductive

---

grounds of exclusionary permission (for example, in a rule-consequentialist justification). However, one might argue that a value-based defence of a permission to disregard the value of potential lives is bound to be unstable, and makes the exclusionary permission only contingently valid. This is true, but also the degree of intuitiveness of the original neutrality intuition can vary in varying global scenarios, where the potential lives might, after all, matter. So, neither the intuition nor the exclusionary permission are meant to be set in stone.

choices. (We want to be able to decide who will be born on the basis of their well-being!)

Broome noted that only the intuition of neutrality allows us to be confident, given that global warming is a bad thing (2005: 405). The exclusionary account allows us to salvage both the intuition and our confidence that global warming is overall bad, that is, something we have overall excellent reasons to fight against. As I have tried to show, the reasons to fight against global warming — its effects on the actual people that do and will suffer from it — would conflict with reasons that are always permissibly disregarded. This difference in normative status is itself an excellent reason to choose to attend to the former.

References

Broome, John. 2010. 'The Most Important Thing About Climate Change', in Jonathan Boston, Andrew Bradstock and David Eng (eds), *Public Policy: Why Ethics Matters*, ANU E Press, Canberra, 101–16.

———. 2009. 'Reply to Rabinowicz', *Philosophical Issues* 19 (*Metaethics*): 412–17.

———. 2005. 'Should We Value Population?', *The Journal of Political Philosophy* 13(4): 399–413.

———. 2004. *Weighing Lives*, Oxford University Press, Oxford.

Lenman, James. 2002. 'On Becoming Extinct', *Pacific Philosophical Quarterly* 83(3): 253–69.

Mulgan, Tim. 2006. *Future People: A Moderate Consequentialist Account of Our Obligations to Future Generations*, Oxford University Press, Oxford.

Narveson, Jan. 1973. 'Moral Problems of Population', *The Monist* 57(1): 62–86.

Rabinowicz, Wlodek. 2009. 'Broome and the Intuition of Neutrality', *Philosophical Issues* 19 (*Metaethics*): 389–411.

Raz, Joseph. 1999. *Practical Reasons and Norms*, Oxford University Press, Oxford.

Part IV
Ramifications

11

Climate Change and Food Justice*

Lori Gruen and *Clement Loo*

Most people paying attention to climate change (CC) have focused on greenhouse gas (GHG) emissions resulting from the use of fossil fuels. Recently, more people are becoming aware that food production, in addition to its distribution, is a major source of greenhouse gas emissions, particularly the practice of raising animals for food. The impact of 'livestock' emissions is so significant that Rajendra Pachauri, the chair of the Intergovernmental Panel on Climate Change (IPCC), made an explicit call urging individuals to 'eat less meat — meat is a very carbon intensive commodity' (Gruen 2011: 89). Less attention has been paid to the climate cost of other aspects of industrial agricultural practices and not enough attention is being paid to predicted food vulnerability due to climate change. The connections between food and climate change are multi-directional and complex, but one thing is clear — feeding ourselves contributes to climate change and our ability to continue to feed ourselves will be significantly affected by a changing climate. Of course, we can't stop eating, but we can evaluate the way food is produced and make personal choices and, more importantly, help shape policies to minimize the impact food consumption has on climate change and that of climate change on access to food. In what follows, we examine the impact of greenhouse gas emissions, arising from industrial agricultural practices, on food insecurity (if climate change predictions are right). We argue that a just solution to the potential problem requires a re-conceptualization of responsibility and urge that strategies to mitigate food insecurity — by

* The term 'food justice' is gaining traction. According to Robert Gottlieb and Anupama Joshi, in a book bearing that title, work on food justice 'seeks to transform where, what and how food is grown, produced, transported, accessed, and eaten' (2010: 5).

industrializing agricultural practices — be re-evaluated as, in all likelihood, they will do more harm than good. Individual consumers as well as corporations and governments, who have gained due to increased food availability resulting from greenhouse gas intensive farming practices or through beneficial changes in climate (or both), have an ethical responsibility to work to aid those who will suffer when food becomes too expensive or unavailable, even if they aren't, in a narrow sense, causally responsible.

Industrial Food Production and Climate Change

Over the last century, food production in the United States (US) has shifted from small family farming operations that relied on human and animal labour to grow a variety of crops, to an intensive industrial food production system. While this transformation allowed for increased yields, the costs for small farmers, non-human animals and the environment have been substantial. The proportion of farmland in the US has remained relatively stable at 40 per cent since 1900, yet American farms went from producing an average of 2 to 3 billion bushels of grain per year in the period leading to the Second World War to producing around 12 billion bushels annually by 2010 (USDA 2011f).[1] A consequence of this greater yield is that industrial agriculture is now responsible for approximately 2 per cent of carbon dioxide (CO_2) emissions and 77 per cent of the nitrous oxide (N_2O) emitted by anthropogenic sources in the US (Wood and Cowie 2004; Johnson et al. 2007).

Industrial agriculture requires substantial chemical inputs, particularly from synthetic fertilizers. The amount of nitrogen fertilizer now used is four times of what it was 50 years ago and this increase has a direct impact on climate change.[2] Of the nearly 12 million metric tons of nitrogen fertilizer spread onto fields in the US, only 30–50 per cent is absorbed by plants. The remainder either leaches into waterways or settles into the soil (Tilman et al. 2002). It is estimated that fertilizers introduced into the soil are responsible for the emission of an annual mean of 150.5 million metric tons of

---

[1] Including corn, sorghum, wheat, oats, barley, and rice.

[2] In 1960, only 2.7 million metric tons of nitrogen fertilizer was spread on American fields and pastures (USDA 2011b). As of 2010, the amount has grown to 11.4 million metric tons a year (ibid.).

CO_2 equivalents (USDA 2011b).[3] The production process to create these fertilizers also contributes to greenhouse gas emissions. 97 per cent of the nitrogen fertilizers used in the US are synthetic fertilizers produced using the Haber-Bosch process (ibid.). This process uses hydrogen extracted from fossil fuels to fix atmospheric nitrogen into ammonia and contributes to greenhouse gas emissions in two ways. First, it requires large volumes of steam, which is energy intensive to produce. Second, the primary by-products of the extraction of hydrogen are two greenhouse gases, carbon monoxide and CO_2. According to the European Fertilizer Manufacturer's Association, 1.29 metric tons of CO_2 are created as a by-product for each metric ton of ammonia produced by the Haber-Bosch process.[4]

With the advent of concentrated animal feeding operations (CAFOs), the numbers of animals raised for food has increased drastically. From 1958 to 1962, for example, there was an average of just under 4 million chickens raised on American farms; by 2010, the number had increased to over 9 billion (USDA 1964, 2011c, 2011e). Vast numbers of ungulates are also being raised for food, and enteric fermentation has, thus, become the second largest source of anthropogenic methane emission after natural gas production. The USDA (2011a) estimates that digestion in the guts of cattle, swine, horses, sheep, and goats in the US is responsible for approximately 140.8 million metric tons of carbon equivalent of methane emissions each year. Animal waste — from managed manure alone — accounts for an additional annual emission of 56 million metric tons of CO_2 equivalent greenhouse gases (ibid.) and, since most of the manure produced by the 90 million cattle in the US is unmanaged (USDA 2011a, 2011c), this figure in all probability does not capture the full impact of livestock waste on emissions. Yet, even with this undercount, if one adds the greenhouse gas contribution from enteric fermentation and livestock waste to that of crop production, agriculture in the US accounts for 357 million

[3] N_2O emissions originating in US agricultural soils exceed the emissions coming from all iron and steel production and natural gas heating systems in the US combined (EPA 2011).

[4] This is only the case when methane is the source of hydrogen; when hydrogen is extracted from coal, the process becomes 20–25 per cent less efficient releasing more CO_2.

metric tons of CO_2 equivalent emissions per year (USDA 2011a).[5] To place this number into perspective, in 2011, fossil fuel combustion for all commercial purposes and all residential purposes in the US generated around 224 and 339 million metric tons of emissions respectively.

Climate Change and Food Insecurity

Projections suggest that, depending on the efficacy of efforts to manage greenhouse gas emissions, anthropogenic climate change will, by 2100, result in an increase of average global temperature of between 1.1°C and 6.4°C (Schmidhuber and Tubiello 2007). Along with this increase in temperature, it is anticipated that there will be significant changes in rainfall patterns. In wetter areas, such as the Mediterranean and Southeast Asia, there will likely be more monsoon rains. In more arid areas, such as Sub-Saharan Africa, West Asia, China, and the Pacific Islands, droughts will probably become progressively worse (ibid.). In general, as global warming progresses, rainfall patterns will become more erratic and less predictable, with the amplitude of fluctuations increasing as one approaches the equator (Dore 2005). These predicted climactic changes will have significant impacts upon food production around the world.

Overall global food production is anticipated to remain relatively stable for at least the first half of the 21st century. In the colder countries, at higher latitudes, climate change may actually result in longer growing seasons and improved rainfall that could potentially increase crop yields (Parry et al. 1999). However, there is near universal agreement that in countries, at lower latitudes, climate change already has, and will continue to have, largely deleterious effects on agricultural productivity and food availability (Rosenzweig and Parry 1994; Parry et al. 1999, 2004; Gregory et al. 2005; Schmidhuber and Tubiello 2007; Lobell et al. 2008).

Historically, in many of the countries in more equatorial latitudes, heat and water stress have been key limiting factors to yields. With the warming and longer droughts that will come with climate change, the heat and water stress limiting crop production in these countries will only worsen (Schmidhuber and Tubiello 2007).

---

[5] With livestock accounting for about 203 Tg CO_2 Eq and crops accounting for 154 Tg CO_2 Eq.

Higher temperatures will shorten growing seasons and disrupt vernalization in many crop plants. More heat tends to accelerate the lifecycles of many plants. Decreases in rainfall when taken together with warming in hotter arid regions will result in decreases in soil moisture and increases in evapotranspiration; both will reduce the water available for crops. Predictions suggest that these effects of climate change, when combined, will most likely result in nearly 9–22 per cent reduction in the global crop yields throughout this century (Parry et al. 2004).

In some regions, particularly those in Africa, yields may drop as much as 50 per cent for essential food crops (Lobell et al. 2008). These reductions in yields are expected to have a large impact on food prices, driving them up as much as 12.5–45 per cent (Parry et al. 1999). This is particularly alarming given that many of the countries that will likely suffer the worst decreases in agricultural output — and hence will become the most reliant on purchasing food from abroad — are also those that are the least able to absorb the cost of increasing food prices. For example, many who live in countries such as Malawi, Ghana, Bangladesh, Vietnam, and Guatemala already spend between 60–80 per cent of their household budget to meet their nutritional needs (FAO 2011). So any substantial reductions in local food production and increases in global food prices have the potential to lead to mass starvation.

High food prices already have an important impact on food security in vulnerable countries. The cost of food currently contributes to the chronic undernourishment of more than 235 million people in Africa and more than 560 million people in Asia (ibid.). With the predicted effect of climate change on yields and food prices, these numbers will only grow. If climate change has as large an impact on food production and prices as most expect, the number of those at risk of hunger will increase by 23 per cent in South Asia, 33 per cent in Africa, and 100 per cent in Latin America (Parry et al. 1999). In the last 100 years, famine has been responsible for approximately 70 million deaths (Devereaux 2000); if the worst effects of climate change on food security are realized, the number of deaths due to starvation in the next 100 years could be much higher.

Food Justice

What should people, who aren't starving, do in the face of great human suffering from hunger, malnutrition and starvation? Of course, before climate change became a matter of concern, people were starving. Large organizations designed to feed hungry children regularly run advertisements on television (and now on the internet) and good people give, not because they believe they have an ethical obligation to do so, but because they see it is a nice act of charity. Feeding those who are hungry is widely thought to be supererogatory — it isn't, strictly speaking, something one has a moral responsibility to do. Some philosophers have argued that this isn't the case (Singer 1972; Pogge 2002; LaFollette 2003).[6] Allowing people to suffer and die from starvation when one can fairly readily do something about it is wrong, especially when those who are starving, such as millions of children who suffer from malnutrition, are in no way responsible for their own hunger. When one is causally responsible for such suffering, not doing something to eliminate it when one can, is, on any account, unethical, some would even say monstrous.

In light of these views, one might argue that because of those of us that in the US and Europe have directly impacted climate change through purchasing food produced by intensive industrial agricultural practices, we should be held responsible for addressing the food insecurity that will arise if predictions are right. This approach is sometimes referred to as 'you broke it, you fix it' and holds that if an individual or collective has taken an action that imposes costs on or harms another, then that individual or collective is responsible for remedying the situation (Moss 2009). It is unjust to make others take on the burdens for one's actions, particularly if the actor has gained and borne little or none of the cost, and the people paying have not benefited at all. Since we all eat, we all are at fault.

However, there is a wrinkle in this view because, while it is true that we all eat, when we purchase the products resulting from intensive agriculture many of us often don't have much choice in the matter.[7] Most consumers in the US and Europe are not directly involved in the intensification of agricultural practices, even though

---

[6] There are differences between the positions that each of these philosophers hold that are not necessarily relevant to this discussion.

[7] In this way, some might say, with Walter Sinnott-Armstrong, 'it's not my fault' (2010: 332). We are encouraging a different understanding of responsibility.

we support it through our purchases and subsidize it with tax dollars. Until recently, even relatively affluent consumers had limited choices about what food to buy (of course, refusing to buy animal products is always an option and will do much to minimize greenhouse gas emissions) and many do not have access to sustainably grown food. While holding those 'who broke it' responsible seems an intuitive principle, and as consumers we all can be said to have contributed to the greenhouse gas emissions generated by the foods we eat, we aren't similarly situated in terms of our food choices. For many people, even in affluent countries, increased greenhouse gas emissions are not the foreseeable outcomes that intentionally result from non-coerced deliberations and actions. Most consumers aren't individually causally responsible for the increase in greenhouse gases that result from intensive food production. At the grocery store, if there is one in the neighbourhood, there aren't products that are 'emission-free' to purchase, consumers don't deliberately choose to purchase high greenhouse gas emitting food. There may be organic or locally produced vegetables available to some consumers who can afford it, and that might minimize emissions, but most individuals don't have that option. Individually, then, while some people can eat more sustainably and cut down or eliminate the consumption of animals, it is initially plausible to believe that we should not hold every individual equally responsible for harmful actions or consequences when each does not intend a harm and made the best choice of those available, even when those choices and actions collectively cause harm.

Perhaps we should place the blame on agribusiness and governments that subsidize it. After all, they are in a position to make choices about industry standards and practices and to change them. Interestingly, intensive agribusiness and those pushing to expand it don't see themselves as the ones who 'broke it' either. Indeed, they think they are trying to 'fix it' where 'it' refers to food insecurity. They are simply responding to demand and working to most efficiently feed a growing population.

In contrast to this causal responsibility model, there is the consequentialist model of responsibility. Here, those in positions to positively affect change have a responsibility to do it, even if they aren't causally responsible for the harm. Wealthy people and nations are in a position to prevent the suffering and starvation that will likely result from climate change and food insecurity and, thus, should commit themselves to not just changing individual consumption habits, but work to promote sustainable food production and distribution, while, at the same time, encouraging a restructuring of social institutions such that they begin to promote more vigorous conceptions of economic, political, social, and ethical responsibility in terms of agricultural production and everything else.

There are, of course, familiar objections to consequentialist conceptions of responsibility. It is too expansive and holds everybody responsible for everything regardless of their intentions or the care they exercise in their deliberations and actions. For consequentialists, nothing is supererogatory. It has also been suggested that consequentialism is unfair because it seems to hold responsible the very people who are the least able to do anything about the problems of high greenhouse gas-emitting foods.[8]

Perhaps, we needn't completely abandon the causal responsibility model and hold everyone, everywhere, accountable for the emissions that result from the food they consume. Even though we all have to eat, some of us have more options available. There are wider and narrower conceptions of causal connections. If we examine our context and role in a causal nexus, as it were, on many occasions some will notice that they are closer to the harm than they initially thought. We can shift from a view of causation as simply a physical chain of events, to a web or a nexus, a view that more accurately describes the complex social, political and economic relations that exist. Of course, one can presumably describe everybody as implicated in the harms that result from all sorts of global problems because of our ever-increasing global interdependencies. However, if it is true that everyone is responsible because of his proximity to a harm in an accurately described causal web, then they are responsible. But, it is likely that different individuals and collectives are situated in different physical, social and economic proximities to those harms, and thus some should be held more responsible than others. Those who end up benefiting through luck but did nothing to deserve it may bear greater responsibility. Those already vulnerable to the food impact of climate change are much farther out in a causal web than those who benefit from a changing climate. The advantage of a more nuanced account of responsibility

---

[8] For general criticisms of consequentialism's 'demandingness', see Mulgan (2001) and Murphy (2000).

is that it locates the point of debate in a more appropriate and realistic place — the issues to be discussed and explored when figuring out who is obliged to respond are causal proximity and wider states of affairs, rather than simply intentions or narrow states of affairs.[9] If individuals and collectives explore their place in the causal web in order to determine when and whether they are responsible for certain harms, they will also have the opportunity and incentive to re-examine what they might successfully do to prevent or ameliorate harms or where pressure should be exerted in order for others to do so and thus have a chance to alter the shape of the web, its various causes and its effects.

Strategies for Addressing the Effects of Climate Change on Food Security

Those who are implicated in the wider causal web have a responsibility to address the effects of climate change on the food security of vulnerable populations. The natural question to ask is: how do they meet this responsibility? There have been a variety of proposals aimed at addressing this question. While these proposals differ in a number of ways, they tend to share a common element: a call for food insecure countries to increase agricultural output through the adoption of intensive farming practices. These practices include greater mechanization/automation, irrigation, specialization and mono-crop production, and use of fertilizers and other agricultural chemicals (Evans 1998; Gregory et al. 2005; Brown and Funk 2008; Lobell et al. 2008, Godfray et al. 2010). This view has become so well accepted as an approach for addressing food insecurity that the Food and Agriculture Organization of the United Nations (FAO) (2007) contends that, in general, intensive agricultural production systems ought to be developed.[10]

---

[9] To some extent this proposal follows Moss' ability to pay model (2009); however, it is distinct in that our suggestion places greater responsibility on proximity to the effect. So, if you have two individuals who are equally able to pay to mitigate food insecurity, but one has benefited more from greenhouse gas-intensive food production or say one is a meat-eater and the other a vegetarian, then, by our account, the one who benefits or eats meat is more responsible. Space does not permit a fuller discussion of this notion.

[10] More specifically, the FAO (2007) suggests that in areas where the opportunity cost of labour is high but land is low, mechanized large-scale

Obviously, there is a problem with this approach to addressing the threats to food security posed by climate change. As we discussed earlier, intensive agricultural practices have significant impacts on climate. In the US, the extensive use of fertilizer and CAFOs makes agricultural production one of the largest sources of greenhouse gas emissions. It seems then that exporting such practices to food vulnerable nations might not be an appropriate means to ameliorate the effects of climate change.

Regardless of climate change adaptation, it is expected that — unless significant steps are taken to curtail fertilizer use — annual N_2O emissions will increase by 68 per cent from current levels by 2050 (Davidson 2012). It is clear that even the current rate of growth in fertilizer use is unsustainable and will result in significant climate effects. If fertilizer intensive agricultural practices are exported to regions where they are not currently being used, this will only accelerate warming and result in climate change being an even more intractable problem. Already, due to the food security strategy that emphasizes production, Asia has become the largest driver of the growth in demand for fertilizers and fertilizer use is projected to increase by 20 per cent in South America and nearly double in Africa between 2010 and 2014 (FAO 2010).

This suggests that simply advocating the adoption of intensive agricultural practices in food insecure areas is not an adequate solution to the threat posed by a changing climate. If agricultural production must be increased to address threats to food security, more thought must be given to how these increases can be made with the smallest possible climate and environmental impact. Further, strategies aimed at bolstering food security must better balance their attention between production, harvesting and distribution. Some estimates suggest that upwards to 40 per cent of the calories grown on the world's farms are lost before they make it onto anyone's table. While 4,600 calories of edible crops are grown

---

cereal production, extensive livestock production, and slash-and-burn systems should be favoured. In areas where the opportunity cost of land is high but labour low, green-revolution type intensification — through increased irrigation, fertilizer and pesticide use — should be encouraged along with intensive livestock production systems. In places where both are high, cash crop production ought to be preferred and only in places where both are low should traditional subsistence farming be encouraged.

per-capita-per-day globally, only about 2800 calories are available for consumption (Lundqvist et al. 2008). The rest is lost to inefficiencies in harvesting, food distribution and use.

In less industrialized countries there are substantial losses due to inadequate infrastructure (ibid.). Without adequate labour and harvesting technology, farmers are unable to harvest their crops before they are lost to pathogens and pests. Without adequate transport, storage facilities and supply-chain planning, they are unable to prevent spoilage after harvest. For example, in India up to 40 per cent of fruits and vegetables and 12 per cent of grains rot either in fields, transport or storage (ibid.). In more industrialized areas, waste becomes the most important source of inefficiency. Within the US, 10–40 per cent — depending on the particular product — of crops are discarded at the farm level and another 26 per cent at the retail level (ibid.).

Growing animals for food is also a source of inefficiency. On average, for every 1,700 calories fed to an animal, one only gets a return of around 500 consumable calories (ibid.). Yet, despite this inefficiency, according to the US Environmental Protection Agency (EPA),[11] about 30 per cent of soybeans, 80 per cent of corn and nearly all of sorghum eventually becomes livestock fodder. This inefficiency, in addition to illustrating a key manner in which food is wasted, has important climate implications as well. The inefficiency with which animals convert feed to meat means that only 15 per cent of the nitrogen introduced into croplands is eventually eaten by humans (Davidson 2012). Much of the remainder enters the atmosphere and becomes a greenhouse gas.

With the potential to achieve such large gains in food availability simply through the reduction of inefficiencies and waste, food insecurity might be minimized by efforts to improve agricultural practices. More sustainable farming methods that, for example, avoid leaving land fallow, restore soil, minimize fertilizer use, and require less energy may also serve as large-scale carbon sinks. Increasing industrial agricultural practices is also a problematic answer to food insecurity as it will raise worries about justice for future generations.

---

[11] See http://www.epa.gov/agriculture/ag101/cropmajor.html (accessed 12 May 2013).

Ultimately, much more thought must be given to how it is possible to grow food without also contributing to climate change. As individuals, we can be more conscientious eaters, that is, by avoiding animal products, learning more about the climate change costs of our food choices and working to increase accessibility of more sustainable, less greenhouse gas emitting food. Socially and politically, given that many of the threats posed by climate change to the food security of vulnerable nations are imminent — most likely manifesting within the next 40 to 70 years — action needs to be taken now to re-conceptualize agricultural practices so that we all may eat.

References

Brown, Molly E. and Christopher C. Funk. 2008. 'Food Security under Climate Change', *Science* 319(5863): 580–81.

Davidson, Eric A. 2012. 'Representative Concentration Pathways and Mitigation Scenarios for Nitrous Oxide', *Environmental Research Letters* 7(2): 1–7.

Devereaux, Stephen. 2000. *Famine in the 20th Century*, IDS working paper no. 105, Institute of Development Studies, Brighton.

Dore, Mohammed H. I. 2005. 'Climate Change and Changes in Global Precipitation Patterns: What Do We Know?', *Environment International* 31(8): 1167–81.

Evans, Lloyd T. 1998. *Feeding the Ten Billion: Plants and Population Growth*, Cambridge University Press, Cambridge.

Food and Agriculture Organization of the United Nations (FAO). 2011. *The State of Food Insecurity in the World: How Does International Price Volatility Affect Domestic Economies and Food Security*, FAO, Rome.

———. 2010. *Current World Fertilizer Trends and Outlook to 2014*, FAO, Rome.

———. 2007. *State of Food and Agriculture*. FAO, Rome.

Godfray, H. Charles, John R. Beddington, Ian R. Crute, Lawrence Haddad et al. 2010. 'Food Security: The Challenge of Feeding 9 Billion People', *Science* 327(812): 812–18.

Gottlieb, Robert and Anupama Joshi. 2010. *Food Justice*, MIT Press, Cambridge.

Gregory, Peter J., John S. I. Ingram and Michael Brklacich. 2005. 'Climate Change and Food Security', *Philosophical Transactions of the Royal Society* 360(1463): 2139–48.

Gruen, Lori. 2011. *Ethics and Animals: An Introduction*, Cambridge University Press, Cambridge.

Johnson, Jane M.-F., Alan J. Franzluebbers, Sharon Lachnicht Weyers, and Donald C. Reicosky. 2007. 'Agricultural Opportunities to Mitigate Greenhouse Gas Emissions', *Environmental Pollution* 150(1): 107–24.

LaFollette, Hugh. 2003. 'World Hunger', in Raymond G. Frey and Christopher H. Wellman (eds), *Blackwell Companion to Applied Ethics*, Blackwell, Oxford, 238–53.

Lobell, David B., Marshall B. Burke, Claudia Tebaldi, Michael D. Mastrandrea et al. 2008. 'Prioritizing Climate Change Adaptation Needs for Food Security in 2030', *Science* 319(5863): 607–10.

Lundqvist, Jan, Charlotte de Fraiture and David Molden. 2008. *Saving Water: From Field to Fork — Curbing Losses and Wastage in the Food Chain*, Stockholm International Water Institute, Stockholm.

Moss, Jeremy. 2009. 'Climate Justice', in J. Moss (ed.), *Climate Change and Social Justice*, Melbourne University Press, Melbourne, 51–67.

Mulgan, Tim. 2001. *The Demands of Consequentialism*, Oxford University Press, Oxford.

Murphy, Liam. 2000. *Moral Demands in Non-Ideal Theory*, Oxford University Press, Oxford.

Parry, Martin L., Cynthia Rosenzweig, Ana Iglesias, Günther Fischer, and Matthew Livermore. 2004. 'Effects of Climate Change on Global Food Production under SRES Emissions and Socio-Economic Scenarios', *Global Environmental Change* 14: 53–67.

———. 1999. 'Climate Change and World Food Security: A New Assessment', *Global Environmental Change* 9 (supplement 1): S51–S67.

Pogge, Thomas. 2002. *World Poverty and Human Rights*, Polity Press, Cambridge.

Rosenzweig, Cynthia and Martin L. Parry. 1994. 'Potential Impact of Climate Change on World Food Supply', *Nature* 367(6459): 133–38.

Schmidhuber, Josef and Francesco N. Tubiello. 2007. 'Global Food Security under Climate Change', *Proceedings of the National Academy of Sciences of the United States of America* 104(50): 19703–708.

Singer, Peter. 1972. 'Famine, Affluence, and Morality', *Philosophy and Public Affairs* 1(1): 229–43.

Sinnott-Armstrong, Walter. 2010. 'It's Not My Fault: Global Warming and Individual Moral Obligations', in Stephen Gardiner et al. (eds), *Climate Ethics: Essential Readings*, Oxford University Press, Oxford, 332–46.

Tilman, David, Kenneth G. Cassman, Pamela A. Matson, Rosamond Naylor, and Stephen Polasky. 2002. 'Agricultural Sustainability and Intensive Production Practices', *Nature* 418(6898): 671–77.

United States Department of Agriculture (USDA). 2011a. *US Agriculture and Forestry Greenhouse Gas Inventory: 1990–2008*, Government Printing Office, Washington.

———. 2011b. *Data Set: Fertilizer Use and Price*, http://www.ers.usda.gov/Data/FertilizerUse/ (accessed 10 May 2013).

———. 2011c. *Overview of US Livestock, Poultry, and Aquaculture Production in 2010 and Statistics on Major Commodities*, Government Printing Office, Washington.

———. 2011d. *Meat Animals Production, Disposition, and Income: 2010 Summary*, Government Printing Office, Washington.

———. 2011e. *Chickens and Eggs: 2010 Summary*, Government Printing Office, Washington.

———. 2011f. *Feed Grains Data: Yearbook Tables*, Government Printing Office, Washington.

———. 1964. *Livestock and Poultry Inventory, January 1: Number, Value, and Classes*, Government Printing Office, Washington.

United States Environmental Protection Agency (EPA). 2011. *Inventory of US Greenhouse Gas Emissions and Sinks: 1990–2009*, Government Printing Office, Washington.

Wood, Sam and Annette Cowie. 2004. 'A Review of Greenhouse Gas Emission Factors for Fertiliser Production', *IEA Bioenergy Task* 38, http://www.ieabioenergy-task38.org/publications/ (accessed 6 May 2013).

12

Climate Refugees: A Case for Protection

Gianfranco Pellegrino

The Problem

By the end of this century the global sea level will rise somewhere between 28 and 43 centimetres as a result of thermal expansion and the melting of glaciers and ice caps (IPCC 2007: 409). Moreover, this rate is not uniform; regional variances can be experienced, with small island states (such as the Polynesian archipelagos Tuvalu and Kiribati) likely to suffer disproportionate consequences especially in terms of land loss (Gillespie 2003–4; IPCC 2007: 413–14; Loughry and McAdam 2008). Also, countries with low-lying coastal areas are under this threat. More than 13 million people across European countries could be affected due to flooding as a result of one metre rise in sea level. Especially vulnerable are coastal regions in the Netherlands, Belgium, Germany, Romania, Poland, and Denmark (see EEA 2006: 22–23). The situation is even worse in regions of high population density, such as South Asia. A rise in sea levels of 45 centimetres would displace 5.5 million people and submerge over 10 per cent of Bangladesh, with increased levels of migration (IPCC 2007: 569). Some regions have been hit by violent climate-induced disasters, such as devastating floods and typhoons. During the 1995 Berlin conference on climate change, Atiq Rahman, of the Bangladesh Centre for Advanced Study, warned the audience: 'if climatic change makes our country uninhabitable, we will march with our wet feet into your living rooms' (Athanasiou and Baer 2002: 23). Many theorists claim that people forcefully displaced by submersion and violent floods should be regarded as refugees of a new kind, to be called 'environmental refugees', and that they are entitled to asylum in other countries (see El-Hinnawi 1985; Cooper 1998; Myers 1997, 2002: Conisbee and Simms 2003; Bell 2004: 135–36; 611–12; Williams 2008; Risse 2009).

This chapter has two parts. First, I will put forward an argument in favour of the claim that people whose displacement is caused by the given climate-induced events could be properly defined 'refugees', in conformity with the official definition given by the 1951 Geneva Convention relating to the Status of Refugees, and treated according to the established legal provisions concerning admittance of political refugees. The argument in support of this claim will be based on the idea that refugees have a specific *right to a territory*, understood as the right to a settled placement within a given territory.[1]

[1] Here I will mention rights in conformity to the usage of this term in current legal and political discussions, rather than as it appear in philosophical debates. In other words, I am not endorsing any strong notion of rights nor any deontological vision of morality. I am only assuming that there are people entitled to a given treatment, in force of the joint support of moral arguments, such as the ones discussed in the next section, and legal documents, such as the 1948 Universal Declaration of Human Rights. In particular, I will not assume that rights are to be considered primitive or not derivative, and the same holds for basic freedoms. Rather, I tend to consider them as claims to various entitlements, to be justified by reference to more basic values, whose achievement is made possible by establishing certain rights (see Griffin 2008); but this view will not be discussed here. Nor I endorse the notion that rights should ever be respected, no matter the costs of this respect. I simply assume that respect for a core set of *basic* human rights — namely, life, property and fair treatment under the rule of law — and freedoms — to wit, freedom of action, association, movement, thought and speech — is *ceteris paribus* worthy of being pursued (on the notion of 'basic rights', see Shue [1996]; notice that here I am endorsing Shue's notion, but not his substantive list of basic rights). The only function I give to right protection is to provide us with a sufficient justification of the state's typical functioning. In particular, I submit that state coercion could be justified only in terms of its function in preventing violations of human rights and undue restrictions of basic freedoms. This view, also, will be posited without a full discussion. (Such a minimal standard for state legitimacy is adopted in Rawls [1999: 1978–81]. See also Buchanan [2010: chs 1, 2 and 5]. Stronger standards, in connection with territorial rights, are advocated in Nine [2012: ch. 2].)

In recent debates, rights to territory are understood in terms of the following meanings: collective rights, giving to certain groups the entitlement to an exclusive jurisdiction on persons and resources on a given territory, including the right to determine residency rights and to control movement

Second, I will argue that environmental refugees and their right to a territory calls in question the *right to a territory of states*, understood as the right that states have to hold *exclusive jurisdiction* within their territory. This claim will be supported through the following two arguments:

(*a*) *The over-population argument:* admittance of environmental refugees might deprive admitting states of the control of their population size, and therefore of a necessary element of exclusive jurisdiction;

(*b*) *The outside interference argument*: environmental refugees and the ensuing loss of exclusive jurisdiction are due to outside collective choices constraining the autonomy of admitting states. This leads to jeopardizing the state's territorial rights, as well.

Environmental Refugees as a Legal Category

The 1951 Definition and its Normative Grounds: Three Interpretations

The definition of 'a refugee', given in the 1951 Convention, establishes that

> the term 'refugee' shall apply to any person who ... owing to well-founded fear of being persecuted for reasons of race, religion, nationality, membership of a particular social group or political opinion, is outside the country of his nationality and is unable, or, owing to such fear, is unwilling to avail himself of the protection of that country; or

---

across borders (see Buchanan 2003; Meisels 2009: ch. 2; Miller 2011; Nine 2012: chs 1 and 3). In this chapter, I am going to assume that the *collective* territorial rights of states or groups are paralleled by an *individual* right to a territory of individuals. I will not discuss and defend this assumption, though. Nor will I explicitly discuss various definitions of the collective territorial rights. Notice, however, that Nine (2012: 9) claims that the 'authority to determine residence, immigration, and citizenship' is a merely 'contingent, not necessary, part of a territorial right'. An implication of the view defended in this chapter, is that states' control on their borders, population and residence can be seriously weakened by admission of climate refugees. I conclude that this puts individual rights to territory in opposition with collective territorial rights. However, if Nine's point is endorsed, this contrast is less relevant.

> who, not having a nationality and being outside the country of his former habitual residence as a result of such events, is unable or, owing to such fear, is unwilling to return to it (art. 1A).[2]

Against the claim that people fleeing from submerged areas could be included in this official definition, the following objection is currently raised: environmental factors of displacement are not covered in the official definition. Hence, those groups of migrants are not properly refugees; at most, they can be defined environmental migrants. Use of the term 'refugee' for this kind of displaced people is inappropriate. Even though environmental causes can be traced back to the government's inaction or negligence, this falls short of the traditional legal notion of persecution that is at the core of the Convention: 'this is very different from situations where government-induced relocation (such as forced relocation due to development projects, displacement resulting from natural hazards, or environmental accidents) may create or contribute to the refugee problem' (Williams 2008: 508). Therefore, any claim to the effect that those migrants should be admitted on the same legal grounds appealed to when political refugees are considered is devoid of legal validity. Let's call this the *definitional challenge* (see ibid.: 508–9).[3]

---

[2] On the 1951 Convention, see Hathaway (2005: 91–110).

[3] Supporters of the claim that environmental refugees constitute a distinctive legal category might face further objections. The most discussed are: a *pragmatical objection*, to the effect that any inflationary extension of the legal treatment due to political refugees to large masses of people could heighten states' resistance to admittance. As a consequence, both environmental-caused migrants and ordinary refugees could see their predicament made worse. Putting in question, or trying to re-negotiate, the 1951 Convention itself may be a dangerous move, since many states view even that minimal platform as prescribing terms too generous, and could exploit re-negotiations to narrow it down, or to repeal it at all (see Dummett 2001: 52; Keane 2004: 214–17; Lister 2013).

A *sceptical objection* could be phrased as: according to the best theories, migration has complex and plural causes, generally distinguished in three sorts (see Kritz et al. 1992): the so-called 'push' factors (socio-economic and political features of the country of origin, such as poverty, high unemployment rates, population growth, and political persecutions and conflicts); the 'pull' factors (socio-economic and political features of the

My purpose in this section is to argue that, according to its best interpretation, the import of this official definition can properly include climate or environmental refugees. The normative underpinnings for the 1951 definition of the status of a refugee could be expounded in at least three different ways. First, it could be argued that this definition attracts our attention to the violation of human rights connected to, and ensuing from, political persecutions. Plainly, any individual on earth has a duty to respect human rights, and to repair for their violation, if she can. Then, it seems that states are specially required to do their best efforts in protecting and respecting human rights, just because they have greater possibilities in doing this than any individual could possibly have. In this respect, asylum is the response that states are required to give to violations of human rights. Let's call this the *human rights interpretation of the refugees convention.*

This interpretation faces the following objection: according to this view, any violation of human rights, or at least any violation ensuing from, or connected to, political persecutions might provide the ground for a request of asylum. This would make any political persecution on individuals, no matter whether performed by states or non-state agencies or even when performed by specific individuals, a reason to grant asylum to the victims. However, the claim that any violation of human rights, irrespective of the author, should be faced by admittance of its victims to another state is

---

target countries, such as high employment rates, higher living standards and permissive entry policies); structural elements (socio-economic features of individual migrants, such as education, availability of money for long journeys, or socio-economic characteristics of the global markets and richer states, and so on). No conclusive arguments have been made to the effect that climate-induced events could be a further, and specific, kind of cause, able to lead to migration independently of the other — more typical — factors (see Kibreab 1997; Black 2001). At most, climate-induced events are merely intensifiers, or enablers, of these usual drivers of migration. Therefore, there are neither environmental refugees, nor environmental migrants. There are only migrants and refugees of the more familiar kinds, who can come from countries experiencing environmental disasters, but have been driven to migration by the ordinary causes. For recent illustrations of the debate, see Morrissey (2012) and Westra (2009). In the main text, I will not consider those objections. I will focus only on the definitional challenge.

plainly overstated. It seems more plausible to claim that violations of human rights committed by individuals and groups should be prevented and repaired by the state that holds a legitimate and exclusive jurisdiction over the territory in which these violations occurred. Accordingly, the victims of those violations might have no entitlement to seek asylum or to flee their territory; on the contrary, they should ask their own states for protection, and not foreign states or the international community at large. Therefore, it seems that the human rights interpretation fails to account for the holders of the duty specified in the 1951 Convention, that is, the international community of states.

A better view might be that refugees are not fleeing simply as a consequence of violations of their human rights; rather, they are to be seen as the fleeing victims of a violation perpetrated *by their own state*. In particular, it might be argued that the state's coercive powers are not warranted or legitimate non-derivatively, but only instrumentally — namely only to the extent that their exercise ensures protection and respect of human rights. The main ground of legitimacy for states might exactly be their role in protecting the human rights of their citizens, and repairing violations of them. Hence, insofar as a state violates the human rights of its citizens, it loses its legitimacy and the victims of its persecution are entitled to look for the necessary state protection elsewhere. Consequently, any legitimate state has the duty to protect not only its citizens, but also any stateless individual.

This view of state legitimacy might have a further consequence. Not only do states have a legitimate role to the extent that they ensure protection and respect of human rights, but also their territorial integrity is instrumental to that purpose. In other words, closed political communities and their mutual isolation through regulated borders, are warranted only since, as a matter of fact, by dividing up this way the territory of the world the maximal respect of individual human rights is ensured. (If, in fact, it turns out that a cosmopolitan institutional framework would be the best way to protect individual human rights, then, according to this view, states would lose their legitimacy.) Consequently, any state to which the victims of state violations of human rights address their requests of protection is obliged to protect them, ceteris paribus. Admittance and asylum, in this respect, constitute a sign of the protection that admitting states are prepared to bestow on refugees. Let's call this the *state protection interpretation of the refugees convention*.

Protection of human rights and basic freedoms is a matter mainly of adjudicating and enforcing law. It seems apparent enough that control over territory is required to this purpose, and it seems evident enough that such a control could be more difficult for a supranationally, or even a globally centralized administration, than for smaller, and less centralized, structures. Clearly, this argument might be used even within states, to claim that also a centralized, big state is unfit to a territorially pervasive defence of human rights. However, this objection could be taken as an argument against centralization and big states, not against states of medium size and with a good deal of subsidiarity. Be this as it may, what matters to the present purpose is that the role of states in protecting against violations of human rights seems empirically grounded, at least given the many failures of supra-national attempts to prevent those violations in many parts of the world.

Against the state protection interpretation of the refugees Convention, the following objection might be raised. Strictly considered, state violations of human rights do not require admittance and asylum. Rather, they are the ground to carry out humanitarian interventions, or (in some cases) to wage a just war.[4] Insofar as the victims of state violations of human rights flee in large groups from their territory, of course, admittance and asylum could be considered required responses. However, as a matter of fact, flight and displacement are strategies open only to a few of the victims of gross state violations of human rights. In many cases, the bulk of the victims is forced to stand in their territory, not being able to move. Accordingly, the relevant response to these cases should be active intervention, rather than a passive behaviour constituted of mere admittance of the few asylum-seekers coming from places where large violations of human rights are perpetrated by states. Admittance of refugees cannot be substituted for active humanitarian intervention.[5]

To overcome this objection, an alternative argument might be proposed in support of the official 1951 definition. Refugees are not

---

[4] On this topic, see Buchanan (2010: part III).

[5] On these topics and the two interpretations of the 1951 Convention, see Garvey (1985), Edwards (2001) and Lamey (2012).

simply the victims of generic violations of human rights. Rather, a specific right is denied to them, that is, *the right to a territory*. Refugees are stateless people in two senses: first, they are no longer protected by their states; second, in order to escape political persecutions, they cannot help abandoning a given territory — typically, their birth-place or the territory where they have settled their living. Now, it can be shown that it is a violation of this specific right that leads to a right to admittance and asylum. To this purpose, I employ the following line of reasoning (borrowed from Dummett 2001: 28, with some modifications).

(*a*) *Citizenship*: Every individual is entitled to live in the country of her citizenship, at least because this is a necessary condition to enjoy state protection (and providing this protection is the only justification of the existence of states, as claimed previously).[6] As a consequence, no state may lawfully banish or expel its own citizens. Indeed, possession and enjoyment of this right to live in one's country is one of the marks of citizenship;

(*b*) *Right to a territory*: If every individual has the right to live in the country of her citizenship, then *a fortiori* she has the right to live somewhere. To put it otherwise, if every person has the right to be protected by her state, she has the right to be protected by a state, whatever it is. If settlement in a territory is a necessary condition for state protection, then the right to live somewhere is a necessary condition to enjoy state protection;[7]

(*c*) *Right to admittance*: If it is unjust for a state to deny its citizens the right to live in its territory, then it is unjust for any state to deny individuals the more general right to a territory. Arguably, people who are denied their right to live in the

[6] This entitlement is acknowledged by the 1948 Universal Declaration (see art. 13).

[7] State protection, as already said, is aimed at preventing and repairing violations of human rights. Accordingly, right to a territory is instrumental to protection of rights of a different kind. This does not make right to a territory less relevant or basic, though. The fact that the justification of a right to a territory is not primitive, or independent, does not undermine its value or point.

country of their citizenship are no longer within the jurisdiction of their own state; indeed, even if only in a figurative sense, their state no longer exists, at least not as a responsible or legitimate state. Accordingly, the other legitimate states have the duty to take in both stateless people and people unjustly expelled from their country of citizenship;

(*d*) *Right to admittance for refugees*: States could expel their citizens also by persecuting them, that is, by making expatriation the only viable option to save one's life and goods. Accordingly, expatriating victims of state political persecutions are entitled to admittance and asylum in other states.

Let's call this interpretation of the normative foundations of the 1951 Convention, the *right to a territory interpretation*. It has many virtues: first, it is able to account for the specific entitlement given to refugees in the 1951 Convention, that is, their right to have a territory to live in; second, through this interpretation, the right to a territory can be easily connected with the foundation of state legitimacy — namely, protection of human rights. However, this connection does not make right to asylum and duties of admittance collapse into duties of humanitarian intervention or international active rescue (as in the state protection interpretation). In other words, the right to a territory interpretation keeps the meaningful normative elements of the state protection interpretation but, unlike the latter, it is able to fit the precise duties and rights established in the 1951 Convention.

Moreover, this account provides us with an easy route to distinguish ordinary migrants from refugees, as well as with a barrier to any inappropriate extension of the category of refugees. The right to a territory interpretation entails that migrants could be labelled as refugees only if two conditions are met. First, they cannot help expatriating; second, they are in this predicament because of state action or of lack of state protection. In other words, refugees are not ordinary migrants, in that the latter, even in the worst situations, have other (if not necessarily better) options. Ordinary causes of migration are not necessary drivers of it, for many people do not migrate even in scenarios where the most commonly acknowledged causes of migration are present. By contrast, for refugees, expatriation constitutes the only viable option to secure their rights and freedoms. In addition, even when migration constitutes a compelling option, this could be the case also when the state made its best efforts to protect its citizens, as it might be the case when acute and sudden economic crises obtain.

Hence, the claim that refugees are entitled to free admittance does not amount to claiming that migrants of any kind should be freely admitted. It might be the case that both claims are true, and yet arguing in favour of admittance of refugees (at least if the 1951 Convention is interpreted as proposed here) does not amount to arguing in favour of open borders. Even though it could be argued that migrations due to serious catastrophes such as sudden economic crises also require admittance, this claim cannot be supported by appealing to the 1951 Convention, or to the right to a territory interpretation of it.

Right to a Territory and Environmental Refugees

The right to a territory interpretation of the 1951 Convention could be employed to show why people displaced in consequence of a complete submersion of their place of living could be properly considered refugees, and should be treated accordingly. In those cases, the distinguishing marks of the refugee status occur. Indeed, those circumstances provide a particularly sharp instance of a state violation of the right to a territory.

People living in coastal areas submerged as a consequence of climate-induced catastrophic events, if not moved in other parts of their country with the help of their state, if not helped by it in other ways, and if not able to move on their own, cannot help expatriating. Besides, the catastrophic events of submergence and flood typically cause the destruction of their material goods, and often put at risk their lives and health. In addition, those kinds of events could be forecasted, with a sufficient reliability, contrary to other sudden and catastrophic climate-induced events. Accordingly, taking provisions for these outcomes is plausibly part of the state's protective duties, such as taking provisions against political persecutions or abstaining from committing them. Hence, the citizens of submerged areas have both their basic human rights and their freedoms threatened. Moreover, the state protection they deserve is lacking, since their state is at fault in failing to take adequate provisions to prevent the catastrophe, or at least it failed in its best efforts to do it. It seems plain enough that those people are entitled to admittance and asylum in other states, on grounds which are strictly analogous to those invoked in the case of ordinary political refugees.

However, environmental refugees display distinctive features. Ordinary political refugees lose their right to a territory because their permanence in their place of living has been made extremely difficult and dangerous by political persecution, or by a lacking state protection. As previously explained, refugees cannot stand in the country of their citizenship since permanence there is too risky for them. In other words, threats to their human rights and basic freedoms force them to give up their right to live in their country.

By contrast, environmental refugees lose their right to a territory in a more straightforward and literal way. They lose their territory, which is literally destroyed by a catastrophe. Accordingly, the violation of their human rights is not the cause of the violation of their right to a territory, but rather the consequence of it. The destructive impact of climate on their territory and the negligent behaviour of their state rob them of the territory they were entitled to.

From this difference a normative difference issues. Ordinary political refugees could claim not only a right to admittance and asylum, but also a right to return home, when state protection of their human rights and basic freedoms will be assured there. Consequently, it might be argued that refugees have only a temporary right to asylum, whereas their strongest claim is to return their country. This makes clear that asylum falls short of full, and permanent, citizenship.

Moreover, admittance of refugees is different from admittance of migrants in another respect. It seems that proper refugees should be freely admitted, but that their admission is conditional and temporary, being dependent on the possible restoration of their right to live in the country of their citizenship. By contrast, while admittance conditions for migrants could be less lenient and more piecemeal than those for refugees, the status of temporary migrants and their rights is controversial (see Ottonelli and Torresi 2010; Bauböck 2011; Lenard and Strahele 2012). After all, the bad economic conditions, which the migrants are escaping, are often permanent, or at least relevantly long-lasting. Moreover, if migrants are admitted in virtue of a concern for the freedom of movement, any temporal limitation would seem an arbitrary limitation on their liberty.

Climate refugees are different. There is no territory for them to come back into. Accordingly, they could claim *a new territory* and permanent citizenship in it. This move is not puzzling, at least according to the right to a territory interpretation of the 1951 Convention; indeed, it is a straightforward entailment of this interpretation.

Territorial Rights of States and Environmental Refugees

Territorial Rights of States

State protection of citizens' human rights and freedoms might be achieved only insofar as the state exercises an *exclusive jurisdiction* within a given closed territory. Exclusive jurisdiction in a territory might in its turn be understood as the right to exercise coercive powers to make, adjudicate and enforce legal provisions within that territory, without interference from outside forces (Nine 2012: 90–92). It seems, in particular, that to have a closed community and a bounded territory is an obvious requirement to be free of interference from outside forces, and therefore territorial determined borders, and a control on them, turn out to be necessary elements of exclusive jurisdiction. After all, national self-determination and national control on the internal social and political life could be shown to be necessary means to exercise individual freedoms and to enjoy basic human rights for the citizens of any state. This right to exclusive territorial jurisdiction might be labelled, for short, *the territorial right* of the state in question. Like the right to a territory for individuals, this right is a contingently necessary consequence of state protection of human rights and basic freedoms.[8]

Then, we have two different rights here, connected to the same normative source, but with a different scope and nature. *Individual* rights to a territory are rights to a settled establishment, rights to

[8] Notice that such a jurisdiction should not necessarily be without limits. It is conceivable that the state jurisdiction is exclusive in certain areas, but subject to external controls in others. For instance, in the European Community, economic policies of states are in many areas controlled and implemented at the community level. However, insofar as national states apply and enforce these decisions and regulations, and governments take part in the community-level legislative and deliberative process, exclusive jurisdiction holds.

reside within a given territory; by contrast, territorial rights *of states* are rights to exclusive jurisdiction within a territory. Individuals gain their right to a territory since this right is a necessary condition for their enjoying state protection of their human rights and freedom. States gain their territorial rights since those are necessary conditions for their granting protection to their citizens and the latter is a condition for their legitimacy (see Stiltz 2009).

Refugees and the Limits of Territorial Rights of States

Territorial rights of states find their roots in the value of human rights and freedom. This value also sets their limits. The treatment of refugees is a clear case of this limiting action of the goal of human rights protection. Refugees have lost the state protection which is due to them, and this in its turn deprives them of the right to reside in their country. Then, this right of theirs should be restored, thereby ensuring the much-needed state protection of their human rights and freedom. An action required to do this, among others, is to grant them the right to reside in another country, at least until the conditions are reached for restoring their original right to reside in their country. Being a relaxation of the state control of borders, this also constitutes a relaxation of, and a limit on, the territorial rights of any state hosting refugees.[9] However, these relaxed territorial rights are not in opposition to the normative underpinning of the notion of a right to a territory, both for individuals and for states. Protection of human rights and freedom warrants both kinds of territorial rights we envisaged and their limits.

The treatment due to climate refugees could be accounted for in this framework too. They should be admitted, in order to have their right to a territory restored. However, as previously remarked, climate refugees cannot come back to their original territories, and therefore their admittance should be permanent, contrary to other kinds of refugees. Due to this fact, the hosting states should limit their territorial rights, not only concerning their control on borders, but also with respect to the control and regulation of their population size. This could be viewed as a further relaxation of their territorial rights. However, this additional relaxation could be somewhat paradoxical.

---

[9] A different approach to a similar claim concerning the limits in scope of the territorial rights is defended in Nine (2012: 39–44, ch. 8).

Territorial rights and Climate Refugees: A Dilemma

Admittance of environmental refugees has the main consequence of altering the population size of the receiving states, especially because their admittance should be permanent. Now, this seemingly neutral fact could seriously undermine the territorial rights of the admitting state. Beyond certain limits of population size, a state's capacity of efficient protection against possible violations of the human rights of its citizens could be impaired. Of course, natural responses to this threat could be subsidiarity and de-centralization. However, it might be argued that a threshold of population size exists beyond which even a de-centralized, imperial-like state loses its control of substantial parts of its territory. In these cases, it seems unavoidable that the portion of territory where the extra-population lives gains its own jurisdictional autonomy. This event could obtain in two different forms: either the new entity formed by the refugees actually secedes from the rest of the admitting state, or it establishes a federative framework, in which it has an exclusive jurisdiction on its territory, but exercises this jurisdiction within a joint structure with the rest of the original state. In both cases, the original state loses its territorial integrity, since it literally loses a part of its territorial jurisdiction. It seems, then, that the right to a territory of large masses of refugees might well be in contrast with the territorial right of the target states. Let's call this the *over-population argument*.[10]

Moreover, as already noticed, climate refugees are a result of certain climate-induced disasters, such as submergence and floods of islands and low-lying coastal areas. Now, climate-induced events are the distant effect of long chains of drivers, whose responsibility is divided among thousands of agents in the world. In particular, current levels of emissions are due to the intertwined present and past conducts of millions of individuals on earth. In collectively bringing about events that are very likely to force certain groups of people to flee their countries, the rest of the world is, in a sense, collectively determining the options that the countries asked for admittance will face. Accordingly, those countries have their situation determined by outside interferences. Therefore, the sheer

---

[10] Obviously enough, this argument might also be used to advocate control on borders concerning admittance of ordinary migrants.

existence of climate refugees seriously undermines state exclusive jurisdiction because it sets external constraints on it. Let's call this the *outside interference argument*.

It seems, then, that the target states of climate asylum-seekers will face the following dilemma: either climate refugees should be denied admittance, or at least they should be denied admittance beyond a given threshold; or territorial rights of states should be declared void. This would make the treatment to be reserved to climate refugees strikingly different from the one given to traditional political refugees. This difference, however, apparently lacks a plausible foundation. Indeed, the same line of reasoning that seemingly supports territorial rights of states — that is, the necessity of protecting people from state violations of human rights — could be invoked against any differentiation of this sort. Accordingly, either a new doctrine on refugees or a new account of the territorial rights of states should be provided.

Conclusion

In this chapter, I tried to argue two claims. First, environmental refugees can be considered as a distinctive legal category, and to them the treatment is due that the 1951 Convention on the Status of Refugees dictates. Second, environmental refugees, being entitled to permanent residence in a territory, could restrict or undermine the territorial jurisdiction of the admitting states. Accordingly, in the case of environmental refugees, territorial rights of individuals are in opposition to territorial rights of states. Hence, territorial rights of states need a new account or a new foundation.

References

Athanasiou, Tom and Paul Baer. 2002. *Dead Heat: Global Justice and Climate Change*, Seven Stories Press, New York.

Bauböck, Rainer. 2011. 'Temporary Migrants, Partial Citizenship and Hypermigration', *Critical Review of International Social and Political Philosophy* 14(5): 665–93.

Bell, Derek R. 2004. 'Environmental Refugees: What Rights? Which Duties?', *Res Publica* 10(2): 135–52.

Black, Richard. 2001. 'Environmental Refugees: Myth or Reality?', Working Paper no. 34, University of Sussex, http://www.unhcr.org/3ae6a0d00.html (accessed 11 May 2013).

Buchanan, Allen. 2010. *Human Rights, Legitimacy, and the Use of Force.* Oxford University Press, Oxford.

———. 2003. 'Boundaries: What Liberalism has to Say', in Margaret Moore and Allen Buchanan (eds), *States, Nations, and Borders: The Ethics of Making Borders*, 231–61.

Conisbee, Molly and Andrew Simms. 2003. *Environmental Refugees: The Case for Recognition*, New Economics Foundation, London.

Cooper, Jessica B. 1998. 'Environmental Refugees: Meeting the Requirements of the Refugee Convention', *New York University Environmental Law Journal* 6(3): 480–529.

Dummett, Michael. 2001. *On Immigration and Refugees*, Routledge, New York.

Edwards, John. 2001. 'Asylum Seekers and Human Rights', *Res Publica* 7(2): 159–82.

El-Hinnawi, Essaam. 1985. *Environmental Refugees*, United Nations Environmental Programme, Nairobi.

European Environment Agency (EEA). 2006. 'Vulnerability and Adaptation to Climate Change in Europe', EEA Technical Report, 7/2005, European Environment Agency, Copenhagen.

Garvey, Jack I. 1985. 'Toward a Reformulation of International Refugee Law', *Harvard International Law Journal* 26(2): 483–500.

Gillespie, Alexander. 2003–4. 'Small Island States in the Face of Climatic Change: The End of the Line in International Environmental Responsibility', *UCLA Journal of Environmental Law and Policy* 22: 107–29.

Griffin, James. 2008. *On Human Rights*, Oxford University Press, Oxford.

Hathaway, James. 2005. *The Rights of Refugees Under International Law*, Cambridge University Press, Cambridge.

Keane, David. 2004. 'The Environmental Causes and Consequences of Migration: A Search for the Meaning of "Environmental Refugees"', *Georgetown International Environmental Law Review* 16(2): 209–23.

Kibreab, G. 1997. 'Environmental Causes and Impact of Refugee Movements: A Critique of the Current Debate', *Disasters* 21(1): 20–38.

Kritz, Mary M., Lin Lean Lim and Hania Zlotnik (eds). 1992. *International Migration Systems: A Global Approach*, Clarendon, Oxford.

Intergovernmental Panel on Climate Change (IPCC). 2007. *Climate Change 2007: The Physical Science Basis*, Cambridge University Press, Cambridge.

Lamey, Andy. 2012. 'A Liberal Theory of Asylum', *Politics, Philosophy and Economics* 11(3): 235–57.

Lenard, Patti Tamara and Christine Straehle. 2012. 'Temporary Labour Migration, Global Redistribution, and Democratic Justice', *Politics, Philosophy and Economics* 11(2): 206–30.

Lister, Matthew. 2013. 'Who are Refugees?', *Law and Philosophy* 32(5): 645–71.

Loughry, Maryanne and Jane McAdam. 2008. 'Kiribati: Relocation and Adaptation', *Forced Migration Review* 31: 51–2.

Meisels, Tamara. 2009. *Territorial Rights*, Springer, Berlin.

Miller, David. 2011. 'Territorial Rights: Concept and Justification', *Political Studies* 60(2): 252–68.

Morrissey, James. 2012. 'Rethinking the "Debate on Environmental Refugees": From "Maximalists and Minimalists" to "Proponents and Critics"', *Journal of Political Ecology* 19: 36–49.

Myers, Norman. 2002. 'Environmental Refugees: A Growing Phenomenon of the 21st Century', *Philosophical Transactions of the Royal Society London* 357(1420): 609–13.

———. 1997. 'Environmental Refugees', *Population and Environment: A Journal of Interdisciplinary Studies* 19(2): 167–82.

Nine, Cara. 2012. *Global Justice and Territory*, Oxford University Press, Oxford.

Ottonelli, Valeria and Tiziana Torresi. 2010. 'Inclusivist Egalitarian Liberalism and Temporary Migration: A Dilemma', *Journal of Political Philosophy* 20(2): 202–24.

Rawls, John. 1999. *The Law of Peoples*, Harvard University Press, Cambridge.

Risse, Mathias. 2009. 'The Right to Relocation: Disappearing Island Nations and Common Ownership of the Earth', *Ethics & International Affairs* 23(3): 281–300.

Shue, Henry. 1996. *Basic Rights: Subsistence, Affluence and US Foreign Policy*, Princeton University Press, Princeton.

Stiltz, Anna. 'Why Do States Have Territorial Rights?', *International Theory* 1(2): 185–213.

Westra, Laura. 2009. *Environmental Justice and the Rights of Ecological Refugees*, Earthscan, London.

Williams, Angela. 2008. 'Turning the Tide: Recognizing Climate Change Refugees in International Law', *Law & Policy* 30(4): 502–29.

13

Ethical Issues for Education and Climate Change

Christopher Schlottmann

Climate change education is education about, and for responding to, climate change. Its purposes have been formulated in various ways, including science literacy (emphasizing literacy of the natural sciences), different forms of interdisciplinary education (integrating social and humanistic disciplines, such as policy and ethics), and social change education (inclusive of teaching students how to solve problems, often prescriptively). In this chapter, I focus on the ethical purpose of climate change education, specifically the ethics of educating for responding to climate change, concerns of advocacy and urgency, and the role of science as it is related to education about climate change.[1]

Ethics Education

The unique ethical challenges of climate change have been detailed elsewhere, including in this volume. Ethics has a role in most forms of climate change education. If the purpose of climate change education is to advance the understanding of climate change as a physical and social phenomenon, then it must include ethical considerations, such as the harm it causes. If the purpose of climate change education is to improve responses to climate change, and those responses involve ethics, then climate change education requires an account of ethics education. If climate change education aims solely to advance an understanding of climate science, then it at least implicitly avoids a central aspect of climate change. Terms such as 'harm', 'risk' and 'dangerous' have an ethical dimension, and discussion of social impacts readily opens up ethical topics.

---

[1] This chapter is conceptual in orientation, and so covers what I see as some major philosophical and ethical issues that arise with education about and for climate change. My purpose is not to survey the existing field of climate change education curricula.

When education has a moral purpose, it raises questions about what moral purpose (and set of values) is being adopted and the permissibility of teaching those values. Questions about the permissibility of teaching values includes which values are being taught, whether they are contested, if appropriate consent is secured, and if it is consistent with various professional, social and personal ethics. The major ethical issues that arise in this context include the risk to agency or autonomy, risk of indoctrination, and accuracy of learned content.

It is relatively uncontested that some instances of guiding students to accept certain beliefs and behaviours are permissible. In civic and democratic education, students are taught to accept certain beliefs and behaviours of civic participation (Gutmann 1999; Brighouse 2005). This is done in such a way that promotes their ability to reflect on these beliefs, but such reflection is not complete (and, in almost any complex area of understanding, can never be complete). Students in democratic society are encouraged to accept certain political norms, but also made aware of them to encourage retroactive reflection and consent. The difference between educating students to accept basic principles of democratic society and accepting a specific moral belief like deep ecology is that (*a*) the former is significantly less contested, and the latter is contested, having multiple plausible alternatives, and (*b*) democratic education aims to empower students to critically analyse the moral view instead of simply accepting it.

If we accept that some forms of climate change education require advancing our understanding of ethics, what form might this take? The answer depends on what purposes climate change education has, and exactly what climate change education is aiming for.

Purposes

The major qualities of education and environmental education can usefully be categorized into their place (formal or informal), practice (pedagogy) and purpose (aims) (Johnson and Mappin 2005: 2–5). This chapter focuses on the purposes or aims of education, in large part, because it is the most philosophical of these categories. Others have been pursuing important work/research on the practice (CAMEL 2011) and place of such education. Climate change education can have a variety of purposes or aims. One central distinction is between education *about* and education *for* a topic or aim, in this case climate change. This is a conceptual distinction; the two categories are not exclusive in practice. A comprehensive climate change curricula can educate about basic climate change literacy and also *for* some sort of response to it.

Education *about* climate change aims to be purely descriptive, and does not have an intentional aim, for example behaviour change. Traditional approaches to science education take this form, as the aim is literacy about a topic, rather than specific actions, values or behaviour changes. Despite avoiding explicit aims, educating *about* something might have unintended outcomes. For example, education about morality (contrasted with education *for* morality) might unintentionally disenchant or confuse students by presenting various strong arguments for seemingly incommensurable positions. Such outcomes might even lead to students learning that morality is completely relative (Baier 1985).

Education *for* climate change aims to include at least some prescriptive dimensions, normally responding to climate change. This can include enabling or empowering students to think or act in such a way that is perceived to lessen the impact of climate change. Education *for* something can include both general (first-order) and specific (second-order) aims.

First- and Second-order Aims

General aims do not prescribe specific, exclusive and contested views. This includes civic education, democratic education or education for critical thinking, which educate for a wide range of specific ethical aims, and exclude certain oppressive ethical aims, but does not require any one in particular. Amy Gutmann and Dennis Thompson label these as the 'second-order' theories of democracies; the characteristics can apply to educational aims as well. We can call them 'second-order' aims (Gutmann and Thompson 2004: 126).

Education *for* climate change can also call for specific aims. This might include educating for civil disobedience, for cost-benefit analysis, for scientific experts determining policies, or for individual behaviour that lessens one's impact. We can call these 'first-order' aims (ibid.). They are specific, have reasonable alternatives, and are not uncontested.

The distinction between first- and second-order commitments helps to distinguish between value-laden education that is permissible and education that is either impermissible or ambiguous.

Conceptually, the distinction is not perfectly clear. As knowledge conditions and social norms change, so does the degree to which knowledge and values are contested and accepted. For example, to teach that climate change science is supported by tens of thousands of peer-reviewed studies is relatively uncontested (in academic and scientific communities), whereas it would have been untrue 50 years ago.

Insofar as climate change education aims to teach values, these values are considerably more difficult to categorize into first- and second-order aims. First, I will discuss the ethical issues that arise when teaching such aims, specifically advocacy, then the ethics of teaching values in climate change education.

The Ethics of Educating for Climate Change

Is educating *for* climate change, meaning educating for solving climate change and its attendant ethical issues, ever permissible? Even if we assume that educating *about* climate change involves no ethics or implicit values, and that climate change education should *not* educate about the ethics (both assumptions that require justification), we still need an account of ethics in education *for* climate change. Such an account would include the advocacy of teaching *for* particular values, views or behaviours.

Advocacy and Permissibility

The proper role of advocacy in education is an actively discussed topic in educational philosophy (Bok 2006), environmental education and ethics (Jickling and Spork 1998; Jickling 2003; Johnson and Mappin 2005).[2] One major concern of advocacy in education is that it detracts from a student's ability to make informed decisions for herself. This restricts their agency, and thereby their ability to flourish. Another concern is simply that their education will be incomplete, not having considered a wider variety of views. Finally, a student who accepts an advocate's position uncritically might later make poor decisions. An advocate's position is often contested, and contested positions are more likely to be incorrect than uncontested positions. A student accepting the contested

[2] Since this is a brief chapter, the discussion of advocacy is short and not as detailed as it would otherwise be.

position might make poorer decisions in their lives than had they learned about alternatives, thereby causing harm to the student.

Much of the ethics of advocacy hinges on whether the aim of education is a first-order or second-order aim. First-order aims pose a problem for advocacy, as the intended aim is to accept a specific belief, idea or fact that has a reasonable alternative and is most likely contested. This would entail not just narrowing down a student's options of what to believe (almost any decision, given the limitations of time and memory, does this), but would do so dramatically. Imagine that a student was taught (and accepted) that the only viable form of governance is a monarchy. They would have a very restricted set of options when considering what form of civic engagement to pursue, thereby shutting off options that might lead to a richer life. They would not know about viable alternatives to monarchies. And finally, their future decisions might be hindered by not knowing about these alternative options. Monarchy is a political example, but one can also imagine religious (educating for specific doctrines) or ethical (educating for specific moral beliefs) examples with the same outcomes. These three are a partial list of the ethical concerns of teaching *for* first-order aims.

Much hinges on the distinction between first- and second-order aims. If the existence or anthropogenic nature of climate change was scientifically contested, something vigorously argued by the contrarian community, then even scientific education about climate change could be considered impermissible according to this argument. Some have made this argument in the American context, in a form that is analogous to 'teach the debate' arguments concerning evolutionary theory (for example, Reardon 2011). Such an argument treats uncontested content as equal to contested content (for example, evolution and creationism as scientific accounts).

Educating for First Order Aims

If we accept the argument discussed, that it is impermissible to educate for first-order ethics aims without strong overriding reasons, would educating for climate change be permissible? This depends on what the first order aims are and if there are overriding reasons. The difference between educating students to accept basic principles of democratic society and accepting a specific moral belief like deep ecology is that the former is significantly less contested, and the latter is contested, having multiple plausible alternatives.

As mentioned, this raises concerns about the permissibility of teaching with these aims. How does it bear on climate change education? Some first-order aims for climate change education include:

(*a*) Assessing costs and benefits of future climate change based on high discount rates (which is actively contested).[3]
(*b*) The belief that cost-benefit analysis, the precautionary principle, or other ethical decision-making procedures are neutral or exclusively correct.
(*c*) Any account of climate change ethics that does not account for future generations (see Revesz and Shahabian 2011).
(*d*) Any account of climate change which holds that it is merely a technical problem.[4]
(*e*) Educating for minimizing our carbon footprint, recycling, eating local food, or other attempts to minimize individual contributions to climate change.[5]

This list is by no means complete, but does touch on some of the central aims in the climate change discourse. I would characterize all of these as first-order aims, in that they are specific, have reasonable alternatives, and are not uncontested. Each of the aims is either conceptually or empirically contested. Since they are first-order aims, it is impermissible to educate *for* them, absent an overriding reason (for example, urgency).

In addition to the argument that educating for first-order aims would limit student options and diminish their life options and agency, I would argue that it fails to prepare students for the modern world in that it teaches that only one particular tool or framework is valuable. This leaves students with a lack of skills for responding to climate change, should any of the other frameworks or tools be more useful. For example, a student intimately familiar with technical solutions to climate change would be at a loss when faced with non-technical challenges.

---

[3] See the Nordhaus/Stern debate in 2007, or Revesz and Livermore (2008).

[4] See Gardiner (2011), for one criticism of this view.

[5] For one example of the difficulty defending such actions, see Maniates (2002).

Educating for Second-order Aims

If educating for climate change with first-order aims is likely impermissible, as argued in the previous section, then what about second-order aims? In liberal education, second-order aims include democracy, critical thinking, and the accurate teaching of history. These are aims that are relatively uncontested (at least in the relevant scholarly communities) both epistemologically (when relevant) and ethically. According to these criteria, they would both be permissible and possible, when required. Critical thinking, for example, is arguably required in liberal education, as it enables students to assess their moral and political views independently.

What qualifies as second-order aims for climate change education? These cannot prescribe specific, exclusive and contested views, but can include ethical views if they are not exclusive or contested. For climate change education, some second-order aims include:

(*a*) A scientific overview of the mechanisms, consequences, probabilities, and uncertainties of climate change.
(*b*) Basic competency in understanding how science is conducted and communicated.
(*c*) Distinguishing between descriptive and prescriptive statements (detailed later).
(*d*) An understanding of the interdisciplinary nature of climate change (and the complexity of interdisciplinarity).
(*e*) An understanding of the major ethical issues raised by climate change and the areas of consensus in traditional and climate ethics.[6]
(*f*) An understanding of the consequences of climate change on social systems, including habitation and migration patterns.
(*g*) An understanding of the environmental consequences of climate change.
(*h*) An understanding of the basic political, technical, nontechnical and social responses to climate change.[7]
(*i*) Basic skills of problem-solving, integration, and collaboration.

---

[6] Educating *about* climate ethics might seem to avoid such an aim becoming a first-order one, but students might implicitly learn that all moral views on a matter are equal, instead of learning about the strengths and weaknesses of each. See Baier (1985) for more details.

[7] I categorize this as second-order since preventing significant harm, and understanding the means to do so, is a relatively uncontested concept.

As second-order aims, these don't seem likely to raise advocacy or other ethical concerns for education directed at them and, therefore, seem (at minimum) permissible.

Permissibility and Requirements

These arguments sketch out what is permissible and impermissible regarding climate change education. Since climate change will cause serious harm to current and future people and nature, there is also the argument that some form of climate change education is not only permitted, but required. It is generally uncontested that we are required to provide an education that prepares students to live a good life in the world they have to inhabit. We incorporate technology into classrooms, emphasize understanding and tolerance of cultures across the globe in curricula, and encourage basic civic and moral virtues. These are responses to defining features of the world we live in.

Since climate change is a defining feature of our current world, and will become increasingly more so as current students grow up, educating *for* climate change (in the second-order sense), in addition to being permissible, might be required, given its urgency and gravity. The International Energy Agency, for instance, recently stated that we have only five years to rapidly reduce carbon emissions before climate change becomes irreversible.[8] Such assessments are complex and open to dispute, but few disagree that a dramatic change in carbon emissions is required in the near future.

Given the urgency of responding to climate change, and the consequences of not doing so, what about educating for climate change in the first-order sense? One can imagine educating for radically low-carbon lifestyles, or prohibiting carbon-intensive activities that promote flourishing, such as artistic creation or travel. One could justify this if the harm of climate change could be lessened by such activity (and one could defend limiting student options so radically). Forced migration and resource shortages are likely to lead to social and political instability, and climate change will greatly increase mortality rates. Preventing catastrophic climate

---

[8] See http://www.guardian.co.uk/environment/2011/nov/09/fossil-fuel-infrastructure-climate-change (accessed 12 May 2013).

change is a very important challenge in the coming decades, after all. In such a situation, one can imagine education that coerces low-impact behaviour or certain political participation that violates or diminishes student agency, but could yield some positive impact for climate mitigation. This comes with significant costs to student agency, however, and would require a much more detailed analysis and political justification to defend its permissibility.

The argument that urgency could justify educating for first-order aims raises many questions. For what purpose would one support, for instance, the condition that all students practice low-carbon lifestyles? An independent argument would be needed for overriding student agency. Despite the gravity of climate change, it is unclear that such a reason exists. Further, there are complexities to proposed first-order aims. Emphasizing personal behaviour, for example, suggests that individual consumer behaviour, adopted in large numbers, can lead to climate change solutions. But climate change is fundamentally a collective problem, heavily mediated through policy and economics, and would require nearly universal behaviour change (at least among those with the highest impact) to measurably matter. To underemphasize this would, at least, implicitly misdirect students.[9] Further, arguments from urgency require constraints on their scope. One recent example is the argument that we urgently need to curb civil rights in the face of terrorist threats. To justify such a stance would require an argument for overriding individual rights and student entitlements. Whether or not climate change education meets this standard depends on (*a*) the gravity of the threat and (*b*) the loss of student agency required. Given the complexity of such cases, the argument for preventing them in the first place becomes strengthened. Avoiding such a conflict or trade-off is clearly preferable to having to make it and lose some significant value in the process.

Objections

However, there are some objections to these arguments, as well as objectives that do not fit cleanly into the categories of first- and second-order aims. Some major objections contain the following: first, that contested knowledge is not a settled concept. A lot of

[9] This is part of a similar problem of individuation (Maniates 2002).

knowledge is dynamic, including in the realm of science, and a lot is neither fully contested nor uncontested.[10] Second, climate is unique in that it is a global phenomenon with (*a*) unclear links between actors and victims, (*b*) a very long time span and (*c*) a global scope of moral considerability. This, at least, points us in the direction of a cosmopolitan ethic or other ethic that accounts for everyone in the world (and future). However, such ethics are still somewhat contested, competing with more locally-oriented ethics, and ethics of immediate relationships (see Williams 1985).

Third, there is an unclear line between first- and second-order aims. Ethics that emphasize sustainability (Curren 2013) or the environment (Varner 1998) are relatively new and in development, and still contested. Fourth, there is genuine uncertainty about some dimensions of climate change. Altering the earth's climate can yield many possible outcomes, many of which we cannot predict. Ethics of uncertainty are different than those covering discrete harms. Fifth, many believe that their first order aims are in fact second order ones, raising questions of the political acceptance of any curricula limiting itself to second-order aims. Finally, educating for climate change might very well be a form of social change education, inclusive of teaching students how to solve problems, often prescriptively. This again is a gray area between first- and second-order aims, and one that likely requires political legitimation before being pursued.

The Interface of Science and Ethics in Climate Change Education

The teaching of the science of climate change raises many of the issues discussed earlier in the chapter. Understanding and responding to climate change requires an understanding of its ethical dimension. The Intergovernmental Panel on Climate Change (IPCC) recognizes this in its 2001 Synthesis Report, *Summary for Policymakers*:

> Natural, technical, and social sciences can provide essential information and evidence needed for decisions on what constitutes 'dangerous anthropogenic interference with the climate system'. At the same time, such decisions are value judgments determined through socio-political

[10] Which moral theory to choose is contested, but not to the degree that all aspects of moral theory are contested.

processes, taking into account considerations such as development, equity, and sustainability, as well as uncertainties and risk (IPCC 2001: 18).

On one hand, this statement seems uncontroversial. Political and social decisions concern the values people have, and the world they want to live in, which are fundamentally ethical concerns. On the other, such an insight is often lost, in part, because we understand climate change through a scientific process, largely communicated by scientists. Putting aside the topics in communication that arise, this raises issues of the relationship between description and prescription (see Boykoff 2011 and Nisbet 2009). There are, at least, two concerns that arise in framing climate change education (and climate change) as a solely scientific topic.

The first is the role of science in prescribing political and social outcomes. James Hansen, for instance, writes that '[t]he science demands a simple rule: Coal use must be prohibited unless and until the emissions can be captured and safely disposed of' (2009: 174). Such a statement suggests that science can tell us what we ought to do, rather than being limited to describing the natural world. Similar statements about the dangerousness of the phenomenon are common in climate change literature. Collapsing the distinction between facts and values potentially blocks out participation from non-scientists. It also restricts the conversation about values that is fundamental to democratic decision-making.[11] While this might be a rhetorical attempt to ground a political judgement in the authority of the natural sciences, it is also a mistaken understanding of natural science and, in addition, carries risks. Using the framework described earlier in the chapter, it implies a first-order aim under the guise of a second-order aim.

The second concern is about the emphasis on the authority of science in social decision-making. The view that scientists serve a special, authoritative role in political discourse about climate change is strongly supported by some experts (for example, Thurston 2007: 62), including Hansen (2009). The natural scientific understanding of climate change is often technical, expert-driven, complex, and

[11] I am not arguing that Hansen necessarily believes this, as he heavily values civic engagement. However, the view that science can prescribe social values could very plausibly lead to such an outcome.

uses a different language than our popular discourse (for example, 'uncertain' suggests strong doubt in popular discourse, but is a technical term in science indicating lack of complete certainty). When such topics enter popular discourse and media, they can change to suggest much more doubt than what actually exists in the science.[12] The public and media conversation about climate change often focuses on its existence and cause, rather than its political, ethical and social implications. The former conversation is not only confusing, but arguably unnecessary. The latter conversation is what most experts believe we need to engage in. In effect, a focus on the science of climate change can be strategically used by contrarians to slow down discussions of how to respond to climate change by shifting the conversation from political solutions to scientific complexities.

Finally, science literacy and advancement of knowledge does not automatically yield an advanced understanding of the interdisciplinary nature of climate change, for instance the risks involved (see Kahan 2011). Despite this, much conversation about climate change is framed as an advancement of scientific literacy.[13] A more integrative, interdisciplinary approach, especially integrative of the ethical dimensions of climate change, would improve this problem. One curricular aim that responds to this concern is discussing fact-value distinctions, the naturalistic fallacy, and the role of science and scientists. The precise role of the natural sciences (and other fields) in climate change education will likely continue to be somewhat contested.

Conclusion

Many of the ethical challenges for climate change education are not covered in this brief chapter. They include the challenges of interdisciplinary education (integrating social and humanistic disciplines like policy and ethics — see CAMEL [2011]), science

[12] At least in some cases, this doubt is sown deliberately. See Oreskes and Conway (2010) for a history of the contrarian movement.

[13] A recent example is http://thehill.com/blogs/e2-wire/e2-wire/172479-top-house-democrat-calls-for-national-climate-change-education-campaign (accessed 12 May 2013), although Al Gore's *An Inconvenient Truth* (2006, dir. Davis Guggenheim) also makes this assumption.

studies (how assessment is conducted, how science is communicated, how uncertainty is conceptualized), and the relationship to environmental and civic education.

Given these arguments, a tentative, incomplete list of areas for emphasis in the ethics aspect of climate change education would include:

(*a*) understanding fact-value distinctions, the naturalistic fallacy and the role of science and scientists in policy;
(*b*) developing the skills of ethical analysis;
(*c*) recognizing that presumptively non-ethical approaches to climate change (for example, cost-benefit analysis and management approaches) have values implicit in them, and that using them is at least implicitly endorsing these values;
(*d*) recognizing that significant ethical issues are present in climate change (Gardiner et al. 2010; Gardiner 2011).

As we spell out the important conceptual and ethical dimensions of climate change, we should keep in mind its educational dimensions, and the importance of advancing understanding of it. Education for climate change must reflect the interdisciplinary complexity of climate change itself, as well as the complex, unavoidable role of ethics in understanding and responding to climate change.

References

Baier, Annette. 1985. *Postures of the Mind: Essays on Mind and Morals*, University of Minnesota Press, Minneapolis.

Bok, Derek. 2006. *Our Underachieving Colleges: A Candid Look at How Much Students Learn and Why They Should Be Learning More*, Princeton University Press, Princeton.

Boykoff, Maxwell. 2011. *Who Speaks for Climate? Making Sense of Media Reporting on Climate Change*, Cambridge University Press, Cambridge.

Brighouse, Harry. 2005. *On Education*, Routledge, Oxford.

Climate Adaptation and Mitigation E-Learning (CAMEL). 2011, http://www.camelclimatechange.org/ (accessed 10 May 2013).

Curren, Randall. 2013. 'Defining Sustainability Ethics', in Michael Boylan (ed.), *Environmental Ethics*, Wiley-Blackwell, Hoboken, NJ, 33–44.

Gardiner, Stephen. 2011. *A Perfect Moral Storm: The Ethical Tragedy of Climate Change*, Oxford University Press, Oxford.

Gardiner, Stephen, Dale Jamieson, Simon Caney, and Henry Shue (eds). 2010. *Climate Ethics: Essential Readings*, Oxford University Press, Oxford.

Gutmann, Amy. 1999. *Democratic Education*, Princeton University Press, Princeton.

Gutmann, Amy and Dennis Thomson. 2004. *Why Deliberative Democracy?*, Princeton University Press, Princeton.

Kahan, Daniel M., Maggic Wittlin, Ellen Peters, Paul Slovic et al. 2011. *The Tragedy of the Risk-Perception Commons: Culture Conflict, Rationality Conflict, and Climate Change*, Working paper no. 230, Yale Law School, http://papers.ssrn.com/sol3/papers.cfm?abstract_id=1871503 (accessed 6 May 2013).

Intergovernmental Panel on Climate Change (IPCC). 2001. *Climate Change 2001: Synthesis Report*, http://www.ipcc.ch/ipccreports/tar/vol4/index.php?idp=7 (accessed 6 May 2013).

Hansen, James. 2009. *Storms of My Grandchildren: The Truth about the Coming Climate Catastrophe and our Last Chance to Save Humanity*, Bloomsbury, New York.

Jickling, Robert and Helen Spork. 1998. 'Environmental Education for the Environment: A Critique', *Environmental Education Research* 4(3): 309–27.

Jickling, Robert. 2003. 'Environmental Education and Environmental Advocacy: Revisited', *Journal of Environmental Education* 34(2): 20–27.

Johnson, Edward, and Michael J. Mappin (eds). 2005. *Environmental Education and Advocacy: Changing Perspectives of Ecology and Education*, Cambridge University Press, Cambridge.

Maniates, Michael. 2002. 'Individualization', in Michael Maniates, Thomas Princen and Ken Conca (eds), *Confronting Consumption*, MIT Press, Cambridge, 43–66.

Nisbet, Matthew. 2009. 'Communicating Climate Change: Why Frames Matter for Public Engagement', *Environment: Science and Policy for Sustainable Development* 51(2): 12–23.

Oreskes, Naomi and Erik Conway. 2010. *Merchants of Doubt: How a Handful of Scientists Obscured the Truth on Issues from Tobacco Smoke to Global Warming*, Bloomsbury, New York.

Reardon, Sara. 2011. 'Climate Change Sparks Battles in Classroom', *Science* 333(6043): 688–89.

Revesz, Richard and Michael Livermore. 2008. *Retaking Rationality: How Cost-Benefit Analysis Can Better Protect the Environment and Our Health*, Oxford University Press, Oxford.

Revesz, Richard and Matthew R. Shahabian. 2011. 'Climate Change and Future Generation', *Southern California Law Review* 84(5): 1097–163.

Thurston, George. 2007. 'Air Pollution, Human Health, Climate Change and You', *Thorax* 62(9): 747–48.

Varner, Gary. 1998. *In Nature's Interests?: Interests, Animal Rights, and Environmental Ethics*, Oxford University Press, Oxford.

Williams, Bernard. 1985. *Ethics and the Limits of Philosophy*, Harvard University Press, Cambridge.

14

The Beauty of Climate Change

Serena Ciccarelli *

Our aesthetic appreciation of nature affects how we behave towards it. Historically, aesthetic considerations have played an important role in the management and conservation of natural areas. The protection of aesthetic interests was one of the main reasons behind the creation of the first national parks at the end of the 19th century (Hargrove 2008). In 1972, United Nations Education, Scientific and Cultural Organization (UNESCO) made it clear that the preservation of nature must serve the purpose, *inter alia*, of maintaining natural beauty throughout the world (UNESCO 1972). International conventions and local decisions concerning landscape management often invoke the aesthetic value of nature as a reason to preserve the environment from destruction and degradation. The political and discursive power of aesthetic considerations within the environmental debate is aptly summed up by Holmes Rolston III: 'Ask people why save the Grand Canyon or the Grand Tetons and the ready answer will be "Because they are beautiful. So grand!"' (2008: 325). The move from the aesthetic appreciation of nature to preservation claims is easily understood if preservation[1] is conceived of as a series of policies to protect nature and eco-systems, regardless of those practical benefits human beings may derive from them.[2] Parsons clarifies that the 'aesthetic

* I thank Marcello Di Paola for his insightful comments and Carlo Chiattelli for his precious help in revising this article. I am also grateful to Sergio Cappucci for our stimulating conversations on geology.

[1] The common differentiation between conservation and preservation is the following: the former calls for protecting areas in the name of human prudence, while the latter aims to protect the integrity of an ecosystem beyond practical human benefits (Parsons 2008).

[2] Following Kant (1997), the standard requirement of aesthetic appreciation is disinterestedness: aesthetic judgement does not derive from the fact the object can serve personal interests.

value ... seems to be just what the preservationist wants: it is a reason to preserve nature even though we gain no practical benefit from doing so, and it is familiar enough to seem intelligible' (Parsons 2008: 102).

Today, the beauty of natural places is increasingly threatened by human activity. The potential impact of climate change is the most recent manifestation of human influence on nature, and a source of concern for the preservation of its beauty. In the last decade, we are witnessing the development of a strategy for managing both natural heritage and other natural places — above all the coastline — from the effect of climate change (Council and the European Parliament 2002; UNESCO 2007). This strategy calls for preserving, among other things, the aesthetic interest of these places. The appeal to the aesthetic value of nature, therefore, is a historically tested device that is still in use. However, the stakes today are much higher, considering the increased human power of disfiguring, destroying landscapes and of modifying the Earth environment on a large scale. In the face of all of this, the revival of scholarly interest in the aesthetic dimensions of nature was to be expected.

The main contention of this chapter is that the occurrence of climate change significantly complicates the debate over environmental aesthetics developed during the last 50 years, and makes all attempts of moving 'from beauty to duty'[3] — that is, from the acknowledgement of aesthetic value to an ethical attitude towards nature, such as preservation — which is extremely hard to accomplish. Some likely outcomes of climate change can be aesthetically acceptable, even 'beautiful', by current standards. By contrast, acting against climate change would imply to deeply affect nature and to interfere with biodiversity. In doing so, the right reference is not the aesthetics of nature, but the aesthetics of the human intervention and the environmental design. This chapter does not investigate this matter nor does it challenge the objectivity or the depth of the current positions in the debate about the aesthetics of nature. Instead, it wishes to tease out the potential of current natural aesthetics arguments for appreciating the beauty that will likely result from climate change. In the process, it is argued, the aesthetic foundation of ecological preservationism is fatally compromised. Furthermore, it is maintained that climate change asks

[3] This is the title of a paper by Rolston III (2008).

us to adopt a 'cosmic' perspective, that is, a standpoint that brings together geological time and the morphology of the whole planet as a whole, not just parts of it. This perspective has significant consequences on aesthetics.

In the first part of the chapter I shall focus on the recent development of the aesthetic theories of nature. In the second part, I underline those claims that attempt to bring aesthetics into line with environmental ethics, and in particular with preservation policies. In the third part, I test current aesthetic parameters on a fictional world brought about by two animated films — *Nausicaä of the Valley of the Wind* by Miyazaki, and *Avatar* by James Cameron — wherein the atmosphere has become lethal to humans and yet nature is beautiful. I then concentrate on some likely real-life aesthetic scenarios, occurring in coastline environments as a result of increased levels of carbon dioxide, in order to highlight some paradoxical results of aesthetic preservationism. Finally, I suggest four strands of research for future scholarship in natural aesthetics that will want to adopt the cosmic perspective.

The Current Aesthetics of Nature

Those who set out to investigate the aesthetics of nature at the end of 20th century were faced with two major difficulties. First, aesthetics categories had been developed over the course of the previous century with reference to art. Second, the ways in which we can become acquainted with and influence the natural world today are more diverse than ever. The perception of the formal qualities of natural places or objects, such as colours and shapes, or the contemplation of the scenic qualities of landscapes, no longer fully account for our current understanding of the aesthetic experience, nor do they capture the manifold ways in which we interact with the natural world. These aesthetic concepts[4] now are out of

[4] Formalism and picturesque are aesthetic approaches developed at the end of 18th century while two similar models, the object and landscape respectively, are used in the 19th century. First, the formalism approach and then the object model guide us to appreciate natural objects in the same way in which we would appreciate a sculpture by focusing on its sensuous and design qualities. In contrast, the picturesque approach and landscape model ascribe preference for scenic view points, which imply framing nature and natural scenarios into bi-dimensional scenes (Carlson 1979).

place both within the recent development of aesthetics (Hepburn 1966) and within the new scientific knowledge of the natural environment (Carlson 1979). In order to overcome these difficulties, over the last 50 years new fields of aesthetics of nature have been investigated, addressing mainly two questions: how nature should be appreciated aesthetically, and whether a correct notion of natural aesthetics can play a role in environmental policy.

The answer to the question: 'How should nature be appreciated aesthetically?' leads to a distinction between two, sometimes overlapping, approaches which can be dubbed cognitive and non-cognitive (Carlson and Berleant 2004). The former argues that the aesthetic appreciation of nature relies on the understanding of its essence, terms and mechanisms: something that can be disclosed by common sense, and more thoroughly by the natural sciences, such as biology, ecology and geology. By contrast, the non-cognitive approaches stress that the beauty of an object is mainly revealed by our relation to it. Hence, what counts is our perception, imagination, emotional arousal, and multisensory experience of nature.

Cognitive approaches emphasize the fundamental role that knowledge plays in the aesthetic experience. The information comes mainly from two sources: cultural traditions that tell stories about places and natural features or scientific knowledge that discloses what nature actually is (Carlson 1979; Saito 1998). The scientific approach, the most developed among cognitive ones, borrows the need for categories from the aesthetics developed in art (Carlson 1979). If we take three emblematic paintings from the beginning of the 19th century to the beginning of the 20th century, such as *Napoleon* by Jacques-Louis David, *Starry Night* by Van Gogh and *Les demoiselles d'Avignon* by Picasso, but we ignore what happened in the art, society and history of the period, we will not fully understand and, consequently, appreciate the different kinds of beauty depicted there. We need the categories of realism, post-impressionism and cubism to perceive and appreciate what exactly we are seeing.

The same should happen with nature. Corals, for example: until the 18th century people believed that they were plants. Today we know that, like plants, they permanently attach themselves to the ocean floor but like animals they hunt as their tentacles capture food from the water. We are also aware of the fact that any single structure of coral is actually made by hundreds to thousands of small coral creatures. Finally, some knowledge of natural science makes us aware that the way in which corals interact with the surrounding eco-system make some of these species the world's oldest living organism. All this information change the way in which we see and aesthetically appreciate corals.

A further logical argument that is sometimes taken is a positive aesthetics claim, which acknowledges the beauty of all things natural. Natural things are valued against intentional objects. What counts is their origin: their history and their evolution, which must be free, or almost free, from human intervention. Today, a natural thing can be more or less natural and consequently more or less beautiful. The quality of 'naturalness' is not absolute, and so are its proper aesthetic properties (Elliot 1982). According to the contemporary defence of positive aesthetics, only science reveals the aesthetic quality of naturalness. Science explains nature and, in so doing, it discloses nature's beauty (Carlson 1984).

One of the main criticisms of this view is the following: sciences make all nature beautiful (Stecker 1997). According to the cognitive approach, the more we know, the more we see and appreciate. But if this is the case, the potential basis of our aesthetic appreciation is enlarged up to the point that every part of nature perceived by trained eyes becomes beautiful. Adopting science as the key element in the disclosure of nature's beauty leaves us without any parameters — both for selecting which science is relevant and for narrowing down the need of our knowledge. The standard response to this charge is that we need at least the kind of knowledge that derives from experience and common sense. Science is important in so far as it changes the perception of the object and this change is brought about by an experience of nature informed by what nature exactly is (Matthews 2002).

Some are unsatisfied with this stand. Without necessarily leaving out the role played by scientific knowledge, they stress other elements as the core of aesthetic appreciation, such as the sense of mysteriousness evoked by nature, our engagement, our emotional response, and our imagination. According to the mystery model, since humans are separated from nature, they cannot know what nature is (Godlovitch 1994). We are bound, both sensually and cognitively, and in perceiving nature we cannot overcome these limits. In order to see nature, we need to frame it into different patterns — such as scientific descriptions — but in doing so we perceive only one single expression of nature and we leave out what nature 'really' is. 'Nature is, for us . . . ultimately alien' (Godlovitch 1994: 19). Since anthropocentric undertakings miss what nature really is, the only aesthetics that makes sense is an acentric one, an 'appreciative incomprehension' that claims grasp, but does not capture, the mystery of nature (ibid.: 26).

In contrast, according to the engagement model, the bridging of the gap between us and the object we appreciate should be the starting point of our aesthetic experience (Berleant 2004). This is true with art — the modern aesthetics of which aims to a 'non confining pattern' — and, above all, with the nature that surrounds us (ibid.: 79). Nature exceeds human beings and this is the reason why any cognitive attempt to capture it is incomplete. What remains is, again, our sense of mystery, but for Berleant this discloses the human need to engage with the natural world. We are urged to a 'sensory immersion' driven by the need to feel continuity with what encloses us (ibid.: 83).

Partly on the same line of reasoning, the arousal model claims that 'being moved' by nature is one important source of our aesthetic experience of it (Carroll 1993). Our emotional reaction to some natural feature, such as our astonishment in front of a majestic waterfall, is an appropriate aesthetic response to it, insofar as our emotions are consistent with the main features of what we appreciate — that is, if the waterfall amazes us with its greatness, it actually is a very large waterfall. According to Carroll, we do not need any particular knowledge about the waterfall: our experiencing the waterfall combines with the resulting coherent feelings. The integrated aesthetics model too contends that the perception of sensorial surface of the environment is the beginning of aesthetic experience (Brady 2003). Nevertheless, the aesthetic experience is an active one and it can be enhanced by our imagination, emotion and knowledge. These factors, particularly imagination, can disclose new qualities of nature and deepen our engagement with it insofar as they lead to an 'appropriate' appreciation. Brady lists the criteria for using imaginings, feelings and thoughts in order to avoid a distorted, trivial or 'overly humanising' use of them (ibid.: 169). Their use must be in line both with the features of the environment we appreciate and with our disinterestedness that is letting nature be what it is, and not an instrument of human desires, needs or curiosities.

Aesthetics and Environmental Policies

With some exceptions, both cognitive and non-cognitive approaches investigate some aspects of aesthetics, such as objectivity, depth and the refusal of a certain kind of anthropocentrism, that are consistent with aesthetic preservationism or with ethical attitudes towards nature. During the last 40 years, some scholars have discussed whether it is possible to prove that — starting from George Edward Moore's claim that one should promote what is good — we should promote the good that the beauty of nature is. Natural beauty is an aesthetic good and its loss would represent the loss of a part of the total good in the world (Hargrove 1989; Thompson 1995). So we have a duty to avoid it.

Sober (1986) defends the aesthetic value of nature as the only consistent philosophical foundation for environmental preservation. Both aesthetics and environmentalism value rarity. They stress the importance of context and, above all, they value something for itself: non-instrumental beauty for aesthetics and the integrity of eco-system beyond human benefits for environmentalism. Eaton (1997) endorses the positive aesthetics claim and discusses the idea that beauty 'requires health' (ibid.: 87). We must value and ensure that aesthetic properties are linked with the naturalness of what we appreciate. Others attach to aesthetic appreciation the role of disclosing, and sometimes, ground a moral attitude towards nature. Even though this approach ultimately recognizes the impossibility to found ethics directly and only on aesthetics, there is a proper room within an environmental ethics for a deep and enlarged concept of aesthetics (Brady 2006; Parsons 2008; Rolston III 2008). Some scholars even start from the requirements of environmentalism, and carve out the most suitable notion of aesthetics to fit them (Carlson 2010). The attempts to combine cognitive and non-cognitive elements, bringing aesthetic notions more in line with environmental ethics, are regarded as successful paths for future aesthetics investigations.

Thus, most of the scholars who founded contemporary environmental aesthetics maintain that getting the aesthetics right can impact on the debate about ethics. Much effort has been put by the different approaches to demonstrate that aesthetic appreciation of nature is not trivial, that it is something more than considering colours, lines and scenic views and it can play some role in environmentalism. The capacity for aesthetics to found other environmental values, as well as its objectivity, is being investigated.[5] The revival of environmental aesthetics is, therefore, also linked with its potential capacity to influence the preservation of nature. But climate change challenges this capacity. This is not because current aesthetics would fail to describe a serious, objective and morally engaged experience when it comes to nature changed by climate. Aesthetics and ethics would not go together, since the first would not counsel acting against global warming. Current aesthetics positions are consistent with climate change being a beautiful thing.

Aesthetics and Climate Change

Consider a national park, an area in which human development is not allowed. From an aesthetic point of view its setting serves two goals: an immediate goal to defend an area considered beautiful at a given time; and a future aesthetic goal, to be achieved by protecting the natural habitat to guard biodiversity and the natural evolution of the place. Thus, preservation policies can preserve natural beauty and let nature create something that is likely to be beautiful in the future. Fantasy movies are examples of how the two goals could be met in the distant future, and in doing so they provide a peculiarly appropriate setting for our reasoning. The following two animated movies emphasize the beauty arising from an imaginary future interaction between human beings and a natural environment changed by climate. In this sense, they are different from the standard last man case where the stake is the value of what is not human and the focus is on natural world in itself (Routley and Routley 1980).

Consider Miyazaki's *Nausicaä of the Valley of the Wind* (1984). Thousands of years after a catastrophe destroyed civilization and natural eco-systems, humans have found a way to survive side by side with aggressive, enormous insects and beautiful toxic jungles. Can a climate, lethal to humans, give rise to something beautiful? This is the question that Nausicaä asks herself while walking with a

---

[5] There are two main ways to argue for objectivism: either in providing a scientific cognitive grounding or, more in line with the aesthetic tradition as Brady (2003) does, by highlighting the inter-subjective basis of judgements of taste.

gas mask on her face and appreciating a flourishing and poisonous forest of enormous plants coloured in green and grey shades and a light white pollen rain, dancing in the wind. Nausicaä's appreciation of the toxic jungle is consistent with several aesthetic models discussed so far. The mystery model is apparent in her 'sense of being outside, of not belonging' (Godlovitch 1994: 28). This is consistent also with the engagement and the arousal models. The cognitive model would work as well: the beauty appreciated by Nausicaä is magnified by her familiarity with that extraordinary eco-system. The very fact that such an eco-system can exist and how it works are part of its extraordinary beauty. Nevertheless, Nausicaä's appreciation could be enhanced by her imaginings, which would be coherent with the eco-system's quality and characterized by disinterestedness. The integrated aesthetics approach, thus, works too.

In the fantasy world we can let climate change happen and the resulting scenarios would be places that today we consider beautiful, probably even sublime. The first criterion of preservation would be turned on its head, then: rather than preserving natural beauty, we should let change happen. Nevertheless, climate change will not have immediate effects; by the time toxic jungles arise, humans would have found a way to adapt and breathe without masks, by pressing their thorax. Consider *Avatar*, a movie by James Cameron (2009), set in a future where the Earth's natural resources have been depleted and humans are mining minerals on another planet, Pandora. In Pandora, whose atmosphere is poisonous to humans, the humanoid natives live in harmony with a flourishing, wild nature. Can man find a way to adapt to climate change and be surrounded by a new, beautiful natural scenario? In Pandora, where humanoids live in a lush and novel environment, the second criterion of preservation would not work. Acting against climate change would not lead to ensuring the natural evolution of the environment, nor to preserving the possibility for something new and beautiful to arise. It would be a different way of affecting nature and biodiversity, for which we need separate aesthetics parameters. We need the aesthetics of human intervention, and possibly look at architecture, engineering, landscapes management, and design.

Similar paradoxical results can be obtained also by considering likely real-life scenarios in coastline development. These areas where land, ocean and atmosphere meet, and 60 per cent of the world's population lives, are particularly vulnerable to climate variability. I make two points here: the first has to do with the decision to preserve endangered, beautiful natural scenarios; the second one deals with the likely, beautiful future scenarios that occur, thanks to climate change. Today, coastline areas are protected from rising sea-levels. This could also require the building of seawalls or artificial coastlines, leading in time to the disappearance of natural beaches. This is a case where human beings step in and 'freeze' the current state of a natural site in an unsustainable way. Man could be tempted to build geo-engineering installations in the oceans. The outcome could be a future, uglier scenario, involving large-scale human intervention. By contrast, the alternative would be to accept a change in the coastline environment, by moving humans into the hinterland and letting the coast become a wet-land. Humans would find it difficult to live in the wet-lands, but this is not relevant from an aesthetic point of view. In the near future, climate zones will probably shift; the most likely consequence of climate change is not the disappearance of landscapes and natural elements, but their moving to parts of the world other than those where we are used to seeing them. This can be aesthetically valuable too.

In the very long term, it cannot be ruled out that new forms of biodiversity will emerge. For example, in the distant future, corals will probably increase as a consequence of a rise in the level of carbon dioxide. The high level of carbon dioxide in the atmosphere will become soluble in rainwater that will, in turn, create carbonic acid in the sea, which is 85 per cent of what corals are made of. In the remote future, we could have an original and beautiful landscape — an amazing coral wood, something similar to the jungles of Pandora or Nausicaä's valley.

Conclusion

In the face of climate change, preservation policies based on aesthetic arguments do not make sense. We cannot act against climate change in the name of natural aesthetics. What is at stake is neither an inescapable relativist position in natural aesthetics nor aesthetics' failure to provide a serious basis for environmental policies. We cannot rule out the possibility that climate change will bring about something that is aesthetically valuable by current parameters in the future. Even in the case in which we aim to protect beautiful nature — as it appears today — against climate change we will fail in our attempt, as the example of protected coastline shows and the following reference to the cosmic perspective makes even more evident. Furthermore, if we wanted to intervene against climate change in the name of aesthetics, we would take into consideration the aesthetics of human intervention more than the aesthetics of nature.

The remote future scenarios, which climate change brings about, force us to adopt a cosmic perspective, or at least a very long-term one. Since changes in climate influence the evolution of the Earth, the temporal reference must be geological time while the spatial reference must be the morphology of the Earth as a whole. I briefly sketch here four topics that can open new strands of research towards the connotation of a cosmic perspective. First, the necessity to investigate further the feasibility of long-term environmental policies such as preservation. Lee (1995) argues that aesthetics and geology have always different, often incompatible, goals. The aspiration underlying aesthetic preservation is to freeze 'beauty for ever' against its geological evolution (ibid.: 219). This is particularly true for climate change: its long-run perspective and the huge natural processes at work with it reveal the extreme difficulty for aesthetic preservation and geological evolution to go hand in hand.

Second, this perspective can be the right place to revise the concept of 'Sublime' or, more precisely, to explore the Sublime frame of mind. Sublime describes the human reaction in the face of greatness, a power potentially lethal for human survival. According to Kant (1997), the power of majestic and terrible natural features shows humans their physical frailty. What we perceive as huge and harmful is subject to constant changes. It is claimed that nowadays nature subjected to human's power hardly arouses the feeling of Sublime. It is clear that technology and science usher in a new era of Sublime. Bodei (2008), for instance, claims that Sublime needs human's view outside their habitat, and he quotes the sidereal spaces.[6] I suggest that today the Sublime applies to future scenarios changed by climate and it is then a frame of mind through which

[6] What is beyond Earth's atmosphere, and concerned with the constellations. Bodei quotes the man on the moon as the first step in this direction.

we can imagine how the Earth's space will change. Human imagination deals with it and science fiction tell us about a world that is post-human. Nausicaä's walk inside the toxic jungle, her superior aesthetic emotion in front of nature that is lethal, might represent the 'negative pleasure' in the face of nature changed by climate, a new frontier of the environmental aesthetics.

Furthermore, consider the positive aesthetics and its proposal for appreciating natural places that have been not touched — or just slightly — by humans. In the distant future, increasing human activities affecting the Earth's natural greenhouse effect will accentuate the anthropogenic elements of climate, which in turn blurs the idea of phenomena on which the influence of men is limited or non-existent. Human-made climate change then heavily modifies the quality of naturalness: it cannot be understood anymore as something free, or almost free, from human intervention but only in terms of the absence of human plan. Finally, consider the cognitive approach, which claims that nature's appreciation must rely on what nature really is. Nevertheless, the reasoning about what we could appreciate and possibly preserve from climate change should focus on what nature will be at the time in which change occurs, something that we ignore and that is influenced by our current decisions. Human imagination will then play an important role. Taking seriously the super long-term prospective of climate change, we have to explore our capacity to make aesthetics judgements based on what nature can become in the future and if and how we aim to aesthetically influence this outcome. For what is at risk in the future is the human lifestyle as we know it, rather than the potential beauty of nature.

References

Brady, Emily. 2006. 'Aesthetic in Practice: Valuing the Natural World', *Environmental Values* 15(3): 277–91.

———. 2003. *Aesthetics of the Natural Environment*, Edinburgh University Press, Edinburgh.

Berleant, Arnold. 2004. *The Aesthetics of Environment*, Temple University Press, Philadelphia.

Bodei, Remo. 2008. *Paesaggi sublimi: gli uomini davanti alla natura selvaggia*, Bompiani, Milano.

Cameron, James. 2009. *Avatar,* Twentieth Century Fox Film Corporation.
Carlson, Allen. 2010. 'Contemporary Environmental Aesthetics and the Requirements of Environmentalism', *Environmental Values* 19(3): 289–314.
———. 1984. 'Nature and Positive Aesthetics', *Environmental Ethics* 6(1): 5–34.
———. 1979. 'Appreciation and the Natural Environment', *The Journal of Aesthetics and Art Criticism* 37(3): 267–76.
Carlson, Allen and Berleant Arnold. 2004. 'Introduction: The Aesthetics of Nature', in Allen Carlson and Arnold Berleant (eds), *The Aesthetics of Natural Environments,* Broadview Press, Peterborough, 11–42.
Carroll, Noël. 1993. 'On Being Moved by Nature: Between Religion and Natural History', in Salim Kemal and Ivan Gaskell (eds), *Landscape, Natural Beauty and the Arts,* Cambridge University Press, Cambridge, 244–66.
Council and the European Parliament. 2002. *Recommendation of the European Parliament and the Relative Council to the Performance of the Management Integrated of the Coastal Zones in Europe,* 2002/413/CE.
David, Jacques-Louis. 1801–5, *Napoleon,* Oil on canvas, 261 cm × 221 cm, Château de Malmaison, Rueil-Malmaison.
Eaton, Marcia M. 1997. 'The Beauty That Requires Health', in Joan I. Nassauer (ed.), *Placing Nature: Culture and Landscape Ecology,* Island Press, Washington, 87–106.
Elliot, Robert. 1982. 'Faking Nature', *Inquiry* 25(1): 81–93.
Godlovitch, Stan. 1994. 'Icebreakers: Environmentalism and Natural Aesthetics', *Journal of Applied Philosophy* 11(1): 15–30.
Hargrove, Eugene. 2008. 'The Historical Foundations of American Environmental Attitudes', in Allen Carlson and Sheila Lintott (eds), *Nature, Aesthetics, and Environmentalism: From Beauty to Duty,* Columbia University Press, New York.
———. 1989. *Foundations of Environmental Ethics,* Prentice Hall, Englewood Cliffs.
Hepburn, Ronald. 1966. 'Contemporary Aesthetics and the Neglect of Natural Beauty', in Bernard Williams and Alan Montefiore (eds), *British Analytical Philosophy,* Routledge & Kegan Paul, London, 191–204.
Kant, Immanuel. 1997. *Critica del giudizio,* trans. by A. Gargiulo, Editori: Laterza, Roma.
Lee, Keekok. 1995. 'Beauty for Ever?', *Environmental Values* 4(3): 213–25.
Matthews, Patricia. 2002. 'Scientific Knowledge and the Aesthetic Appreciation of Nature', *The Journal of Aesthetics and Art Criticism* 60(1): 37–48.
Miyazaki, Hayao. 1984. *Nausicaä of the Valley of the Wind,* Isao Takahata/Top Craft.

Parsons, Glenn. 2008. *Aesthetics and Nature*, Continuum Press, London.
Picasso, Pablo. 1907. *Les demoiselles d'Avignon*, Oil on canvas, 243.9 cm × 233.7 cm, Museum of Modern Art, New York. (Acquired through the Lillie P. Bliss Bequest.)
Rolston III, Holmes. 2008. 'From Beauty to Duty: Aesthetics of Nature and Environmental Ethics', in Allen Carlson and Sheila Lintott (eds), *Nature, Aesthetics, and Environmentalism: From Beauty to Duty*, Columbia University Press, New York.
Routley, Richard and Val Routley. 1980. 'Human Chauvinism and Environmental Ethics', in Don S. Mannison, Michael A. McRobbie and Richard Routley (eds), *Environmental Philosophy*, AUN Press, Canberra, 96–189.
Saito, Yuriko. 1998. 'Appreciating Nature on Its Own Terms', *Environmental Ethics* 20(2): 135–49.
Sober, Elliott. 1986. 'Philosophical Problems for Environmentalism', in Bryan G. Norton (ed.), *The Preservation of Species: The Value of Biological Diversity*, Princeton University Press, Princeton, 173–94.
Stecker, Robert. 1997. 'The Correct and the Appropriate in the Appreciation of Nature', *British Journal of Aesthetics* 37(4): 393–402.
Thompson, Janna. 1995. 'Aesthetics and the Value of Nature', *Environmental Ethics* 17(3): 291–305.
United Nations Education, Scientific and Cultural Organization (UNESCO). 2007. *Policy Document on the Impact of Climate Change on World Heritage Properties*.
———. 1972. *Convention Concerning the Protection of the World Cultural and Natural Heritage*.
van Gogh, Vincent. 1889. *Starry Night*, Oil on canvas, 73.7 cm × 92.1 cm, Museum of Modern Art, New York.

About the Editors

Marcello Di Paola is Research and Teaching Fellow, Centre for Ethics and Global Politics, Luiss University, and Assistant Academic Dean, CEA Global Campus, Rome. He has published a book in Italian on gardens and environmental philosophy. His articles have appeared in international journals, such as *Environmental Values* and the *Journal for the Study of Religion, Nature, and Culture*. He has recently guest-edited a special section on Global Justice for the *Global Policy Journal,* and another on the ethics of climate change for *Philosophy and Public Issues*.

Gianfranco Pellegrino is Assistant Professor, Department of Political Science, Centre for Ethics and Global Justice, Luiss University, Rome. He is also Honorary Fellow at the Bentham Project, University College London. He has published two books in Italian, on Bentham and sufficientarian justice, respectively, and several articles on global justice and related issues. He is the executive editor of *Philosophy and Public Issues*.

Notes on Contributors

Serena Ciccarelli is Research Associate, Center for Ethics and Global Politics, Luiss University, Rome, and Consultant on Social Cohesion for the President of the Republic of Italy.

Lori Gruen is Professor of Philosophy, Feminist, Gender, and Sexuality Studies, and Environmental Studies at Wesleyan University, Connecticut.

Joyeeta Gupta is Professor of Environment and Development in the Global South, University of Amsterdam, and Professor of Law and Policy in Water Resources and Environment, University of Delft.

David Held is Master of University College, Durham, and Professor of Politics and International Relations, University of Durham, as well as Professor of Global Justice at LUISS, Rome.

Dale Jamieson is Director of the Environmental Studies Program and the Center for Bioethics, Professor of Environmental Studies and Philosophy, and Affiliated Professor of Law, New York University.

Clement Loo is Andrew W. Mellon Postdoctoral Fellow, College of the Environment, Wesleyan University, Connecticut.

Darrel Moellendorf is Professor of International Political Theory, Goethe University, Frankfurt/Main.

Tim Mulgan is Professor of Philosophy, University of Auckland, and Professor of Moral and Political Philosophy, University of St Andrews, Scotland.

Francesco Orsi is Senior Research Fellow in Philosophy, University of Tartu.

Pragati Sahni is Assistant Professor, Department of Philosophy, Arts Faculty, University of Delhi.

Ronald Sandler is Associate Professor, Department of Philosophy and Religion, and Director, Ethics Institute, Northeastern University, Massachusetts.

Christopher Schlottmann is Clinical Assistant Professor of Environmental Studies, Associated Director of Environmental Studies, and Affiliated Professor of Bioethics, New York University.

Jussi Suikkanen is Lecturer in Philosophy, University of Birmingham.

Index